Tamer Khattab

Optical Code Division Multiplexing

Tamer Khattab

Optical Code Division Multiplexing

A Mechanism for Sub-Wavelength Switching in All- Optical Networks

VDM Verlag Dr. Müller

Impressum/Imprint (nur für Deutschland/ only for Germany)

Bibliografische Information der Deutschen Nationalbibliothek: Die Deutsche Nationalbibliothek verzeichnet diese Publikation in der Deutschen Nationalbibliografie; detaillierte bibliografische Daten sind im Internet über http://dnb.d-nb.de abrufbar.
Alle in diesem Buch genannten Marken und Produktnamen unterliegen warenzeichen-, marken- oder patentrechtlichem Schutz bzw. sind Warenzeichen oder eingetragene Warenzeichen der jeweiligen Inhaber. Die Wiedergabe von Marken, Produktnamen, Gebrauchsnamen, Handelsnamen, Warenbezeichnungen u.s.w. in diesem Werk berechtigt auch ohne besondere Kennzeichnung nicht zu der Annahme, dass solche Namen im Sinne der Warenzeichen- und Markenschutzgesetzgebung als frei zu betrachten wären und daher von jedermann benutzt werden dürften.

Coverbild: www.purestockx.com

Verlag: VDM Verlag Dr. Müller Aktiengesellschaft & Co. KG
Dudweiler Landstr. 99, 66123 Saarbrücken, Deutschland
Telefon +49 681 9100-698, Telefax +49 681 9100-988, Email: info@vdm-verlag.de
Zugl.: Vancouver, University of British Columbia, Diss., 2007

Herstellung in Deutschland:
Schaltungsdienst Lange o.H.G., Berlin
Books on Demand GmbH, Norderstedt
Reha GmbH, Saarbrücken
Amazon Distribution GmbH, Leipzig
ISBN: 978-3-639-09321-6

Imprint (only for USA, GB)

Bibliographic information published by the Deutsche Nationalbibliothek: The Deutsche Nationalbibliothek lists this publication in the Deutsche Nationalbibliografie; detailed bibliographic data are available in the Internet at http://dnb.d-nb.de.
Any brand names and product names mentioned in this book are subject to trademark, brand or patent protection and are trademarks or registered trademarks of their respective holders. The use of brand names, product names, common names, trade names, product descriptions etc. even without a particular marking in this works is in no way to be construed to mean that such names may be regarded as unrestricted in respect of trademark and brand protection legislation and could thus be used by anyone.

Cover image: www.purestockx.com

Publisher:
VDM Verlag Dr. Müller Aktiengesellschaft & Co. KG
Dudweiler Landstr. 99, 66123 Saarbrücken, Germany
Phone +49 681 9100-698, Fax +49 681 9100-988, Email: info@vdm-publishing.com

Copyright © 2008 by the author and VDM Verlag Dr. Müller Aktiengesellschaft & Co. KG and licensors
All rights reserved. Saarbrücken 2008

Printed in the U.S.A.
Printed in the U.K. by (see last page)
ISBN: 978-3-639-09321-6

Abstract

Optical Code Division Multiplexing (OCDMA) is a method used to enable simultaneous transmission of multiple optical data flows over the same fiber using the same wavelength. In OCDMA, isolation between different data flows is achieved using a set of uncorrelated, or loosely correlated, spreading codes to encode the transmitted signal and decode it at the receiver side. The process of encoding and decoding is performed entirely in the optical domain without the need for optical-to-electrical-to-optical conversion. This increases the granularity of traffic isolation in the optical domain while maintaining higher speed switching because of the all-optical encoding/decoding capability. Although code division multiplexing is a well established technique in wireless transmission where all processing of data and switching are performed electronically, there are many challenges in applying this scheme in the optical domain mainly due to the different characteristics of the medium and the fact that negative-valued signals are not easy to produce.

This thesis has three main objectives: to deploy OCDMA as a switching mechanism at the sub-wavelength level in order to increase the granularity of traffic isolation in all-optical core switching, to design new mechanisms that enhance the performance of OCDMA as a multiplexing method over long-haul optical fiber transmissions, and to model the performance of OCDMA based switching and multiplexing mechanisms.

All-optical switching at the core of the network provides very high speed switching. However, it suffers from low utilization or lack of quality of service guarantees due to lack of fine granularity traffic isolation. This thesis presents an optical network architecture called Optical Code Labeled Generalized Multi-Protocol Label Switching (OC-GMPLS), which utilizes OCDMA as a switching mechanism in backbone GMPLS networks. OC-GMPLS uses OCDMA as an all-optical labeling space in GMPLS switching in order to achieve finer granularity switching at the all-optical network core.

The deployment of OC-GMPLS networks mandates performance modeling to show its advantages and to enable tuning of the new network parameters so that performance can be optimized. In this thesis we present an analytical model for the throughput and switching capacity of OC-GMPLS networks. Using our model, we show how to find optimal operating points for OC-GMPLS networks based on physical layer and network layer parameters.

The performance of OC-GMPLS networks depends on the performance of OCDMA transmission, which is affected by the modulation method and the optical spreading codes properties. In order to enhance the performance of OC-GMPLS networks, we take two different approaches. The first approach is based on proposing a modulation mechanism that enhances the communication reliability while maintaining low bit error rate for OCDMA transmissions. Our Chip-Level Modulated Binary Pulse Position Modulation (CLM-BPPM) scheme provides a simple to implement (in the all-optical domain) yet a very powerful physical layer method for sending multiple optical flows using OCDMA while maintaining the Bit Error Rate (BER) due to Multiple Access Interference (MAI) effects between these flows at a low level of about 10^{-12} for 10 simultaneous users. Our method provides a better capability in terms of clock recovery and user activity detection while achieving error rates in the range of those provided by On-Off Keying (OOK).

Performance of OCDMA transmission depends to a great extent on the efficiency of the codes used to perform the multiplexing. In order to tackle this side, we investigate the problem of Optical Orthogonal Code (OOC) design by proposing a method called Rejected Delays Reuse (RDR) for constructing OOCs using an element-by-element based greedy algorithm. We show that our method provides a computationally less complex algorithm for designing OOCs, which makes it more practical. Our analysis and simulation results show that OOCs designed using the RDR greedy method are also higher in multiplexing efficiency than OOCs designed using classical element-by-element constructions. This is because RDR designed OOCs possesses smaller code lengths for the same code cardinality and weight than their counterpart classical element-by-element greedy designed codes.

Contents

Abstract   ii

Contents   v

List of Figures   x

List of Abbreviations   xiii

Acknowledgements   xvi

Dedication   xix

1 Introduction   1

   1.1   Motivation   1

   1.2   Basic Concepts   3

        1.2.1   Photonic Networks   3

        1.2.2   Generalized Multi-Protocol Label Switching   7

   1.3   The Thesis   8

        1.3.1   Objectives   8

1.3.2 Contributions 9

1.3.3 Road Map 11

2 Overview of Fiber-Optic CDMA 13

2.1 Introduction 14

2.2 OCDMA Digital Modulation Methods 15

2.2.1 ON-OFF Keying (OOK) OCDMA 16

2.2.2 Pulse Position Modulation (PPM) OCDMA 17

2.3 OCDMA Spreading Methods 18

2.3.1 Time Spreading 18

2.3.2 Spectral Spreading 19

2.3.3 Hybrid Wavelength-Time Spreading 20

2.4 Optical Spreading Codes (OSC) 21

2.4.1 Optical Prime Codes 21

2.4.2 Optical Orthogonal Codes 22

2.5 OCDMA Decoding Methods 23

2.5.1 Correlation Detection 23

2.5.2 Chip-Level Detection 23

2.6 Encoding/Decoding of Time Spread OCDMA 24

2.7 Concluding Remarks 31

3 OC-GMPLS Core Sub-Wavelength Switching Architecture 33

3.1 Introduction 34

3.2 Code Switch Capable (CSC) Layer 35

3.3 End-to-End Network Architecture 39

3.4 Core Switch Architecture 41

3.5 Edge Switch Architecture 44

3.6 Switching Capacity and Label Space Expansion Ratio 46

3.7 Numerical Results 52

3.8 Concluding Remarks 57

4 Throughput Analysis for OC-GMPLS Core Switching 59

4.1 Introduction 60

4.2 OC-GMPLS Physical Layer 61

4.3 Bit Error Rate for OOK OCDMA 63

4.4 Core Switch Throughput 69

    4.4.1 Pure DWDM System Throughput 70

    4.4.2 OCDMA System Throughput 71

4.5 Numerical Results 72

4.6 Concluding Remarks 89

5 Chip-Level Modulated BPPM Fiber-Optic CDMA 91

5.1 Introduction 92

5.2 Proposed Chip-Level Modulated BPPM (CLM-BPPM) 94

5.2.1 Modulation Method 94

5.2.2 Transceiver Architecture 97

5.2.3 Detection Methods 99

5.2.4 Chip Conflict Resolution Methods 102

5.3 MAI Channel Model 103

5.4 Performance Analysis 106

5.4.1 BER Mathematical Analysis Framework 108

5.4.2 BER for Absolutely Timed Detection with '1'-Value First 115

5.4.3 Optimal Choice of Parameters 118

5.5 Numerical Results 119

5.6 Concluding Remarks 130

6 OOC Construction Using Rejected Delays Reuse 133

6.1 Introduction 134

6.2 Definitions and Mathematical Preliminaries 135

6.3 Classical Element-by-Element Greedy Method for Constructing OOCs . . 142

6.4 Proposed Greedy Algorithm with Rejected Delays Reuse (RDR) 147

6.5 Sub-Wavelength Switching Capacity and Code Efficiency Analysis 152

6.6 Numerical Results 156

6.7 Concluding Remarks 160

7 Conclusions and Future Research 161

7.1   Summary 161

7.2   Major Contributions 164

7.3   Future Work 165

Bibliography 169

A Proof of $p_r \leq 1$ in Equation 4.1 183

B Derivation of Equation 5.28 185

C Asymptotic Computation Time for the RDR Algorithm 188

List of Figures

2.1 High level architecture of an OCDMA transmission system. 15

2.2 Spectral amplitude spreading using diffraction grating. 19

2.3 Spectral wavelength scramble spreading using arrays of FBGs. 20

2.4 Encoding and decoding of time spread fiber optic OCDMA signals. . . . 26

3.1 Label mapping layers in OC-GMPLS. 37

3.2 OC-GMPLS network architecture and end-to-end operation. 39

3.3 Architecture of OC-GMPLS sub-wavelength core switch. 42

3.4 Architecture of OC-GMPLS sub-wavelength edge switch. 45

3.5 Operational regions and load lines for OCDMA systems. 54

3.6 Finding the optimum label space expansion ratio for code length = 901. . 54

3.7 Optimum label space expansion ratio. 55

3.8 3-D plot for number of orthogonal codes. 55

3.9 3-D plot for number of orthogonal codes in error free region. 56

4.1 Encoding and decoding of optical code labels. 61

4.2 3-D plot of BER of OOK OCDMA system. 77

4.3   BER operational regions for OOK OCDMA transmission ($X = 1024$ bits).   78

4.4   3-D plot of optimum BER.   79

4.5   Throughput of single wavelength OCDMA vs. DWDM system.   80

4.6   3-D plot of OC-GPMLS core switch throughput.   81

4.7   Effect of the no. of sub-wavelength flows on throughput.   82

4.8   Effect of code weight on the throughput of OC-GMPLS core switching. .   83

4.9   3-D plot of throughput with code length at optimum BER.   84

4.10 3-D plot of throughput with packet length at optimum BER.   85

4.11 Max. throughput trajectory against no. of flows.   86

4.12 Max. throughput trajectory against code weight.   87

4.13 3-D trajectories of max. throughput.   88

5.1   CLM-BPPM modulation method.   96

5.2   CLM-BPPM OCDMA transceiver architecture.   98

5.3   Illustration of different detection methods.   98

5.4   Illustration of different chip conflict resolution methods.   101

5.5   General CLM-BPPM OCDMA channel model.   104

5.6   Ideal CLM-BPPM OCDMA channel model.   106

5.7   Comparison of BER from mathematical analysis and simulation.   122

5.8   Effect of changing code length on BER.   123

5.9   Effect of increasing number of users on BER.   124

5.10 Effect of changing code weight on BER.   125

5.11 Effect of changing the optical threshold on BER ($N_u = 9$). 126

5.12 Effect of increasing average user activity P_A on BER ($k = 7$). 127

5.13 BER for a system using OOC with optimum code length. 128

5.14 Optimum code length as a function of code weight and number of users. . 129

6.1 Examples of OOC designs using RDR. 152

6.2 Code expansion efficiency factor 157

6.3 Code expansion efficiency factor under error free conditions 158

6.4 Code length under error free conditions 159

List of Abbreviations

AWG	Arrayed Waveguide Grating
BER	Bit Error Rate
BPPM	Binary PPM
CAT	Containerization with Aggregation-Timeout
CCAMP	Common Control and Measurement Plane
CDMA	Code Division Multiple Access
CLM-BPPM	Chip-Level Modulated BPPM
CoS	Class of Service
CSC	Code Switch Capable
DB	Data Burst
DWDM	Dense Wavelength Division Multiplexing
E/O	Electrical to Optical
EDFA	Erbium Doped Fiber Amplifier
ELSR	Edge LSR
ETE	End-to-End
ExOCL	Explicit Optical Code Labeling

FBG	Fiber Bragg Grating
FEC(s)	Forwarding Equivalence Class(es)
FO-CDMA	Fiber Optic CDMA
FSC	Fiber Switch Capable
FSK	Frequency Shift Keying
FTDL(s)	Fiber Tapped Delay Line(s)
GMPLS	Generalized MPLS
I/P	Input
IETF	Internet Engineering Task Force
ImOCL	Implicit Optical Code Labeling
IP	Internet Protocol
ITU	International Telecommunication Union
L2SC	Layer 2 Switch Capable
LED	Light Emitting Diode
LSC	Lambda (Wavelength) Switch Capable
LSP	Label Switched Path
LSR	Label Switched Router
LTI	Linear Time Invariant
MAI	Multiple Access Interference
MEMS	Micro Electro Mechanical Systems
MPLS	Multi-Protocol Label Switching

O/E	Optical to Electrical
O/P	Output
OBS	Optical Burst Switching
OC-GMPLS	Optical Code Labeled GMPLS
OCDMA	Optical CDMA
OCL	Optical Code Label
ODLC	Optical Delay Lines Correlator
OOC	Optical Orthogonal Code
OOK	On-Off Keying
OPC	Optical Prime Code
OSC	Optical Spreading Code
OXC	Optical Cross-Connect
PPM	Pulse Position Modulation
PSC	Packet Switch Capable
PSK	Phase Shift Keying
QoS	Quality of Service
RDR	Rejected Delays Reuse
TSC	Time-slot Switch Capable
VPN	Virtual Private Network
WDM	Wavelength Division Multiplexing

Acknowledgements

I start by stating that I am obediently thankful to God, praise and glory be to him alone, for his countless blessings and mercy. Without his guidance and mercy I can do nothing.

I would like to express my deepest gratitude to my supervisor, Prof. Hussein Al-nuweiri, for his guidance, technical advice, invaluable feedback, understanding, generous support, and friendship. I have passed through many personal and difficult circumstances during the period of my Ph.D. program. Without Prof. Alnuweiri's genuine kindness, I do not think you would be reading these words today.

I am very grateful to Prof. Fayez Gebali, University of Victoria for the valuable time he dedicated for me to discuss ideas on optical code designs.

I would like to thank Prof. Nicholas Jaeger for teaching me the fundamentals of optical devices and for being always helpful and supportive.

I would like to express my greatest appreciation to my thesis committee for their valuable comments, which enhanced the presentation of this thesis.

I would like to thank Professors Will Evans, Nicholas Jaeger, Vikram Krishnamurthy, Robert Schober, Lutz Lampe, Samir Kallel, Zinovy Reichstein, and Vincent Wong for guiding me through the courses I attended with them.

I would like to express my gratitude to my friend Dr. Watheq El-Kharashi for carefully reviewing my thesis. His valuable comments and suggestions have been very useful in enhancing the presentation of this thesis.

I also would like to thank my friends and colleagues Ayman Kaheel, Amr Mohamed, and Maged Elkashlan for the collaborative work we had at the early years of my program.

I have been blessed to come in contact with many wonderful people throughout my study period at UBC. This page would not be enough to mention all of them, but, the least I can say to them is: thank you for making my stay at UBC such a pleasant experience.

Among the many friends that I had during my study period, the following friends had the most positive effect on my life: A. Elharbi, A. Ismail, Abdelgader, Abdelhamid, Amr Mohamed, Amr Wassal, Anwar, Awad, Ayman, Hesham, Hossam, Ihab, Junaid, Khaled, M. Allam, M. Halfawy, M. Rasheed, M. Rehan, M. Senousy, Maged, Mahmoud, Raef, Samer, Tariq, Uthman, Wajeeh, Watheq, Wael, and Yahia.

Any words I say will not be enough to describe my feelings toward my parents. I can only pray to God to reward them for their constant support, persistent encouragement, and most of all for their unwavering and sincere love and care.

I would like to express my love and deep appreciation to the person who stood beside me through all times, my wife Abeer. She sacrificed a lot to support me throughout our life together. She was always there for me with comforting words, encouraging thoughts and a heart filled with love and sincerity.

I would like to extend my thanks to my sister Taghreed and my brother Hany, for being a source of inspiration and prayers.

I would like also to thank my parents-inlaw for all their prayers, encouragement, and support.

I would like to extend my thanks also to my brothers-inlaw and sisters-inlaw for their prayers and encouragement.

Last, but not least, I would like to thank my dear sons, Hossam and Ziad, for making my life full of joy and happiness even during the toughest times.

Dedication

To my beloved mother and father.

"O'God bestow your blessings on them for raising me."

إلى أمي الحبيبة و إلى أبي الحبيب
﴿ رَبِّ ارْحَمْهُمَا كَمَا رَبَّيَانِي صَغِيرًا ﴾

To my dearest wife and lovely children.

"O'God protect them."

Chapter 1

Introduction

In this chapter, we provide motivations for the work done in this thesis as well as some basic concepts related to optical switching and optical code division multiple access methods. We also provide a list of the objectives that this thesis tries to achieve and the contributions made throughout the thesis to meet these objectives. The chapter concludes by a section that describes the flow of the thesis with a briefing of the work done in each chapter.

1.1 Motivation

As the demand for higher bandwidth keeps on increasing from year to year, it is widely accepted that the network backbone needs to be implemented in all-optical switching to accommodate the tremendous bandwidth requirements [1]. Currently, many carriers have already deployed optical fiber cables over most of their backbone networks to allow for higher data rate transmission. However, the new bottleneck for all-optical service-provider backbones will be due to the *slower* electronic switching in the optical network routers and switches. The use of electrical devices in optical networks requires performing optical-to-electrical and electrical-to-optical conversions of packet data at the input and output ports of the routing/switching device, respectively. The conversion to electrical signal is required to allow electronic processing of packet data to extract its switching and routing information. Removing the *electronic bottleneck* requires the use of *all-optical network devices* called *photonic network devices*. Although photonic devices with primitive fiber and wavelength switching functionality [2,3] are being offered today, the development of photonic switches or routers with more complex functionality poses a number of serious design and performance challenges.

Another major challenge to the photonic networking paradigm is the tradeoff between traffic isolation and network optical bandwidth utilization. While optical fiber cables with optical transmitters and receivers allow for very high data rate point-to-point connections at the physical layer [4,5], the problem remains to be how to utilize this bandwidth efficiently while providing enough traffic isolation between different traffic classes. Traffic isolation is particularly essential for Virtual Private Network (VPN) applications, where concerns over data privacy and security prohibit the use of aggregation at the backbone [6–9]. The solution to this problem is non trivial since photonic switching devices lack the very basic elements of their electronic counterparts, such as the ability to store packets in local buffers and process packet headers in the optical domain. The lack of buffering and packet header processing ability in photonic devices means that a significant body of switching strategies that rely on buffer management and packet header processing are rendered useless for photonic networks. Hence, the need for a fundamentally different paradigm for providing service differentiation guarantees in photonic networks.

1.2 Basic Concepts

This section provides a brief description for some of the basic concepts that are necessary to understand the operation of optical networks and optical code division multiplexing techniques.

1.2.1 Photonic Networks

The setup of optical networks where links are optical channels with all switching and routing devices are electrical devices is referred to as First Generation Optical Networks [10]. In such networks, the data exists in the optical domain only while being transmitted from one point to another over the fiber cables.

The use of electrical devices in optical networks adds the requirement of performing optical-to-electrical conversion of data at the input of the routing/switching device in order to be able to process the data to extract its switching and routing information. At the output of the routing/switching devices an electrical-to-optical conversion of the data is needed in order to be able to carry it over the optical link. The process of optical-to-electrical conversion and electrical-to-optical conversion introduces a considerable bandwidth limitation because the devices used for this operation cannot keep up with speeds supported by optical Wavelength Division Multiplexing (WDM) links. It also increases the costs of switching devices, because high speed optoelectronic devices contribute a major factor to device cost [11].

In order to increase the speed of routing/switching devices, the devices must be able to process the packet data and relay it to the right path without a need to convert the data into the electrical form. In other words, these devices must be able to perform optical processing of the packet header information and able to perform optical relaying of the packet data over the correct path from its source to its destination. Using these devices, we can build a network where data always remains in the optical format through its entire path from the source to the destination. Such networks are called *photonic networks*.

In order to allow for data multiplexing (potentially switching) in a photonic manner, packet header processing is replaced by addressing using physical layer properties that can be processed photonically without a need for optical-to-electrical-to-optical conversion. Three major physical layer properties allow for this paradigm: fiber (space) multiplexing, wavelength (frequency) division multiplexing, and optical code division multiplexing.

Fiber Multiplexing

In fiber multiplexing techniques, data are addressed or labeled based on its spacial properties, which are defined by the fiber it is transmitted over. Switching devices relay data from an input fiber to an output fiber based only on the input and output fibers without any data processing. With current technologies, devices called Micro Electro Mechanical

Systems (MEMS) can perform this function and are used for switching in some optical core networks.

MEMS switching technology is based on tiny mechanically moving mirrors that are used to change the light path direction, thus enabling spacial switching of an incoming light from a certain fiber (source) to a preselected output fiber (destination). An example of devices that incorporate this capability and are used at the core of optical networks are Optical Cross-Connect (OXC) switches [12].

Wavelength Division Multiplexing

Wavelength division multiplexing is a well known mature technology in optical networks. In its basic form, WDM is a technique used to increase the carrying capacity of optical fiber cables. This is achieved through multiplexing several optical signals together into one fiber with each one of them transmitting over a different wavelength. Such multiplexing is possible in a photonic fashion using devices called optical wavelength multiplexers. These devices operate using the concept of optical coupling, where two optical signals are mixed together when the fibers carrying them are brought very near to each other for a certain length.

Switching using wavelength has been studied extensively in the literature [13] and incorporated in several switching devices. OXC devices mentioned in the previous subsection are also capable of performing wavelength switching. OXC devices are capable of differentiating data streams based on the input fiber (spatial) and the wavelength of the data stream. By having a configuration table within each device, a cross connection

can be setup between a certain wavelength on an input fiber and an output wavelength

(might be different or the same as the input wavelength) on an output fiber. The group

of these cross connections between different fibers (and wavelengths) form the *light path*

that the data can take from the source to the destination.

Optical Code Division Multiple Access

The concept of using Fiber-Optic Code Division Multiple Access (FO-CDMA)[1] as a

photonic multiple access method started in the mid 80's by the pioneering work of Pruc-

nal *et al* [14] followed by the work of Salehi *et al* [15]. Around the same time, the

development of a family of spreading codes, called Optical Orthogonal Codes (OOC),

that can be used for performing OCDMA was initiated by the work of Chung *et al* [16].

Since then, there have been several publications in the literature addressing different as-

pects of the subject. It was not till the late 90's when research work in this area started

to pick up again, this time with several prototype implementations [17–22].

OCDMA uses spreading codes with high autocorrelation and low crosscorrelation

properties to multiplex several optical data streams on the same fiber and wavelength in

a photonic manner. The multiplexing is performed by spreading different streams using

different codes then recovering (de-spreading) the streams at the receiver using the same

codes.

OCDMA has gained a lot of attention in recent years from the research community

---

[1]Also referred to in this thesis as Optical Code Division Multiple Access (OCDMA)

because of its ability to perform photonic multiplexing at the sub-wavelength level, which enables higher multiplexing (and potentially switching) capacity and finer granularity of flow isolation. Analysis and design of OCDMA sub-wavelength switched systems will be the focus of this thesis.

1.2.2 Generalized Multi-Protocol Label Switching

Multi-Protocol Label Switching (MPLS) is a technique for associating labels to packets to identify their path across the network. Theses paths are called Label Switched Paths (LSPs). LSPs essentially form end-to-end *virtual tunnels* between edge routers. A single network link normally carries labeled packets from multiple LSPs. However, because of the MPLS labels, packets transmitted over a shared link are always distinguishable. By using packet classifiers or filters, packets can be routed, queued, and treated differently based on their label (or LSP number). Essentially, MPLS requires recognition of the packet boundaries in order to extract the label information before switching the packet.

GMPLS is a multipurpose control-plane paradigm that was proposed by the Common Control and Measurement Plane (CCAMP) working group of the Internet Engineering Task Force (IETF) to allow MPLS switching without need to recognize packet boundaries or header information. One of the main differences between MPLS and GMPLS is in the way labels are represented. In MPLS, the label is generally a small header added to the data unit (packet) in a predefined location. In GMPLS, the labeling concept is extended to include the use of physical layer properties.

The application of GMPLS in optical networks can affect the future of IP-over-WDM networks resulting in reducing the number of layers, which can also reduce cost, complexity, and processing overhead in optical backbones [23, 24]. In this thesis, we exploit the generalized switching architecture of GMPLS to introduce a framework for sub-wavelength switching using OCDMA techniques.

1.3 The Thesis

In this section, we discuss the objectives of this thesis, the contributions made in order to achieve the objectives, and a road map for the flow of the thesis work chapters.

1.3.1 Objectives

This thesis has three main objectives:

1. The first objective is to enhance the switching capacity (granularity) at the sub-wavelength level in all-optical (photonic) networks using OCDMA techniques. The unique features of the OCDMA technique, such as the ability to perform multiplexing and demultiplexing entirely in the optical domain and the ability to work at the sub-wavelength level makes it a potential candidate for the objective. At the same time, the properties of optical transmission such as the lack of bipolar signalling and the use of power as information carrier pose challenges to applying previously developed electrical based CDMA methods to optical networks.

2. The second objective of this thesis is to enhance the multiplexing capacity and transmission properties of OCDMA through designing new spreading codes construction methods and new OCDMA modulation methods that maintain low bit error rates. The existence of multiple access interference in OCDMA methods poses a challenge on developing methods that minimize this effect while maintaining good bandwidth utilization.

3. The third objective is to develop mathematical models for analyzing the performance of OCDMA as a multiplexing technique as well as the performance of optical code based sub-wavelength switching mechanisms. The trade-off between increasing switching capacity in terms of number of multiplexed flows and the increase in packet error rates due to increased multiple access interference makes performance modeling an essential tool for tuning the network performance.

1.3.2 Contributions

The major contributions of this work can be highlighted as follows:

- The proposal of a new OCDMA-based sub-wavelength switching architecture in all-optical networks called Optical Code Labelled GMPLS (OC-GMPLS). The proposed architecture enhances the all-optical switching granularity of core optical GMPLS switches by a large factor. The results for this appear in [25, 26].

- The derivation of an analytical model for calculating the OC-GMPLS switch throughput. The developed analytical model takes into consideration the effect of both the network parameters as well as the physical layer parameters. This work appears in [26, 27].

- The proposal of a new transmission method based on binary pulse position modulation for OCDMA. The proposed method enables the receiver to differentiate between idle and zero data transmissions within one bit duration while maintaining an error rate within the range of On-Off Keying (OOK) systems. This work appears in [28].

- The establishment of a general mathematical framework for calculating the bit error rate due to multiple access interference in OCDMA systems based on time domain spreading with optical orthogonal codes and correlator receivers. This work appears in [28].

- The development of a new algorithm to construct optical orthogonal codes for sub-wavelength switching. The developed algorithm has lower computational complexity and provides shorter code lengths under the same code constraints, which increases the label space utilization. This work appears in [29, 30]

Some of the other minor contributions are:

- A survey of OCDMA communication systems. This survey provides a discussion on the state-of-the-art methods used to modulate, spread, and detect OCDMA signals. It also provides a discussion on the different spreading codes used for OCDMA transmissions. This survey appears in [31].

- A mathematical design for optical delay lines correlator encoders/decoders used for spreading and de-spreading OCDMA signals in the time domain. This work appears also in [31]

- Introducing the concept of *label space expansion ratio* as a measure for the change in flow isolation granularity due to sub-wavelength switching and *code expansion efficiency factor* as a measure for the closeness of a non-optimal OOC to an optimal OOC in terms of label space expansion capabilities. This work appears in [25, 26, 29, 30].

- The derivation of an optimal operating point for achieving maximum label space expansion ratio for OC-GMPLS core switches and the demonstration through numerical results that there is an optimal operating point (not necessarily the same as the former) for maximizing the switch throughput. This work appears in [26].

1.3.3 Road Map

Chapter 2 gives an overview of OCDMA communication systems. It describes the high level transceiver architecture, with discussions on the different modulation, spreading and detection methods as well as spreading code families deployed in the system. The chapter then provides a detailed description of the operation as well as a mathematical-based design of time spread passive correlator encoder/decoder devices.

Chapter 3 proposes an optical sub-wavelenth switching architecture based on introducing a code switch capable layer in GMPLS architecture that uses OCDMA for labeling. The architecture is called Optical Code Labeled GMPLS (OC-GMPLS).

Chapter 4 presents an analytical model for calculating the throughput and switching capacity of OC-GMPLS core switches.

Chapter 5 proposes a new OCDMA modulation scheme based on PPM called Chip-Level Modulated Binary PPM (CLM-BPPM). A general BER framework is established to analyze the BER of the scheme.

Chapter 6 presents a new less complex algorithm for constructing shorter optical orthogonal codes. The algorithm is called Greedy Algorithm with Rejected Delays Reuses (RDR).

Chapter 7 concludes this thesis and outlines some future research ideas.

Chapter 2

Overview of Fiber-Optic CDMA

In this chapter, we provide a background on the necessary components and methods used to enable OCDMA system compilation. We will also provide a detailed mathematical design for the Optical Delay Line Correlator (ODLC), which is a device used to enable all optical spreading in the time domain. The flow of the rest of this chapter is as follows. A brief introduction is provided in Section 2.1. In Section 2.2, we discuss the digital modulation methods used for OCDMA. Following that, Section 2.3 discusses the different optical spreading methods with an emphasis on time domain spreading, which will be the method used in the rest of this thesis due to its simplicity and practicality of implementation. In Section 2.4, we introduce the different spreading code families used in OCDMA with more details on Optical Orthogonal Codes (OOC) given in Chapter 6. Section 2.5 focuses on the optical decoding methods used to de-spread the optical signal. Following that, in Section 2.6, we provide a detailed mathematical design for ODLC encoders and decoders used in this thesis as spreading and de-spreading devices. We finish the chapter by concluding remarks in Section 2.7.

2.1 Introduction

Since its start, OCDMA has been considered for use in optical fiber communications as a multiple access (multiplexing) technique, especially in all-optical Local Area Networks (LAN) [32–34] due to its flexibility and feasible implementation in the all-optical (photonic) domain. Recently, the use of this multiplexing technique has been proposed at the Wide Area Networking level [35], particularly in Multi Protocol Label Switching (MPLS) [36] and Generalized MPLS (GMPLS) networks [25].

A FO-CDMA system uses a set of spreading codes called Optical Spreading Codes (OSC) with certain autocorrelation and crosscorrelation properties to multiplex several users at the same time on the same wavelength over a fiber-optic communication channel. A high level block diagram of an OCDMA system is shown in Figure 2.1. The transmitter generally requires a digital modulator to convert data from its digital format into some analog format that can be carried over the lightwave. The output of the modulator controls a light source (or laser source) that carries the data over lightwave carriers. The modulated lightwaves are then fed into a photonic OCDMA encoder, which performs the optical spreading of the signal. The spread optical signal is sent over the fiber channel. It must be noted that the order of these blocks can change if spreading is performed in the electrical domain, in which case modulation and spreading will be performed in the electrical domain followed by the light source, which will be controlled by the spread signal.

At the receiver side, the reverse is performed. De-spreading of the optical signal is

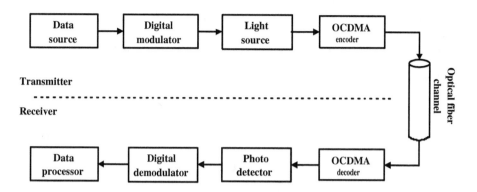

Figure 2.1: High level architecture of an OCDMA transmission system.

performed. The output signal is fed to an optical detector (photodetector) that converts the signal to the electrical domain where it is demodulated and processed by the receiver.

2.2 OCDMA Digital Modulation Methods

With current technologies, it is practically infeasible to devise light sources (transmitters) such as lasers and light emitting diodes (LEDs) that produce a monochromatic (single carrier frequency or wavelength) signals [37]. It is also very hard to control the phase of the light carrier as well as its polarization while it travels through long distances of fibers due to the geometrical and material imperfections of optical fibers [2, 38, 39]. Moreover, integrated light detectors (photo detectors) are generally capable of detecting the power of the incident light signal rather than its frequency or polarization [40]. Due to these limitations, it is more practical to carry the information over the power of the optical signal (corresponding to modifying the amplitude) instead of its phase, frequency, or polarization. Although optical transmission systems that use Phase Shift Keying (PSK) [41,42] and Frequency Shift Keying (FSK) [3,43] modulations have been studied in the literature, amplitude modulation is still the scheme of choice for optical transmission systems due to its ease of implementation at the integrated optoelectronics level. For this reason, we will focus our discussion in this section on modulation schemes that are based on carrying the information using the signal amplitude.

2.2.1 ON-OFF Keying (OOK) OCDMA

Most optical transmission systems use the power of the optical signal rather than its frequency or phase to carry information. This is why OOK was the first candidate for digital optical modulation. OOK is based on carrying the '0' and '1' information bits on the lightwave by switching the light source OFF and ON respectively. This means that a '1' is represented by an optical pulse (light), and a '0' is represented by no pulse (dark). To spread the signal in the time domain, the bit duration T_b is divided into n chip locations (time slots) each with duration $T_c = T_b/n$, where n is the length of the spreading code. A '1' is represented by the transmission of k pulses (chips) at chip locations defined by the spreading code within a one bit duration T_b. On the other hand, a '0' causes nothing to be sent (dark). This scheme is easy to implement and has less requirements in terms of synchronization between transmitter and receiver. However, clock recovery and differentiation between *active sending '0'-bit* and *idle* source states is very difficult to perform at the receiver side.

2.2.2 Pulse Position Modulation (PPM) OCDMA

To overcome some of the problems of OOK, PPM was used as a modulation scheme [44–52]. Classical PPM systems divide the bit duration T_b into M symbol slots, each with duration $T_s = T_b/M$. A pulse at symbol position number i represents the transmission of a symbol with value i out of the M-valued possible symbols. In order to spread this signal in the time domain, each symbol slot is divided into n-chip slots each with duration $T_c = T_s/n = T_b/(Mn)$. Symbol i is represented by a series of k pulses at chip locations defined by the spreading code within symbol slot i. This means that the PPM modulation is applied at the bit level, then the transmitted encoded pulse is spread using the optical time spreading code. This maintains the correlation properties of the spreading code within one symbol duration T_s only while it is violated outside the symbol duration. The disadvantages of bit level PPM are: (i) More strict requirements on chip synchronization (accuracy less than $T_c/2$) and (ii) The possibility of interference between different symbols, because they are similar at the chip level (i.e. if we don't consider the symbol position within the bit). In order to maintain the correlation over the entire bit duration, we propose in Chapter 5 a scheme that uses the bit value to perform PPM of the spreading pulses (at the chip pulse level) rather than at the bit duration level. Maintaining the correlation properties means that we can treat a stream of bits as a cyclic repetition of the spreading code, which is what OOC is designed for.

2.3 OCDMA Spreading Methods

Code multiplexing in fiber-optic communications can be performed using either time domain spreading [53] or spectral (frequency/wavelength) domain spreading [54]. In addition hybrid schemes that use time domain spreading and frequency (wavelength) domain hopping were also discussed in the literature [55, 56].

2.3.1 Time Spreading

In time spreading methods, light sources generate pulses that represent the transmitted data. The pulse duration (called here the chip duration T_c) is usually very small compared to the data bit duration T_b. Replicas of the pulse with reduced energy are repeated at pseudo randomly selected locations (slots) within the bit duration causing a spread of the pulse energy. It is the function of the receiver, which knows the pseudo random spreading pattern (spreading code), to recombine the energy of the pulse replicas so that their energies add and a detection of the original data is performed. This recombination can be performed using either active devices such as a light source at the receiver that correlates only with time slots corresponding to the pseudo random code or passive devices such as an Optical Delay Line Correlator ODLC. ODLCs [15] are devices based on optical fiber tapped delay lines to function as optical correlation filters. ODLCs are utilized by optical transmitters and receivers to encode (spread) and decode (despread) the optical signal. ODLC detailed design and functionality will be described in Section 2.6.

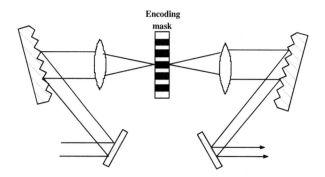

Figure 2.2: Spectral amplitude spreading using diffraction grating.

2.3.2 Spectral Spreading

In spectral spreading methods, a pulse from a light source is spread in the frequency domain by passing it through a device that separates its wavelength components and scales the spectral amplitude of each wavelength component differently or rearranges (scrambles) their order in time according to a pseudo random spreading code.

Spectral amplitude encoding [57, 58] is performed using diffraction gratings as shown in Figure 2.2. Light which is incident on the diffraction grating is separated into its wavelength components. These components pass through a mask that scales the different wavelength components differently according to the spreading pseudo random code. The resulting scaled wavelengths are recombined again to form the encoded signal, which is sent to the receiver. The receiver reverses the scaling operation to recover the original signal.

Wavelength scrambling [59] is performed using an equally spaced array of fiber Bragg

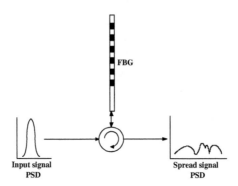

Figure 2.3: Spectral wavelength scramble spreading using arrays of FBGs.

gratings as shown in Figure 2.3 or arrayed waveguide gratings. The gratings are designed

so that each one of them reflects a certain wavelength. The pattern of the grating array

is chosen according to the spreading code. When light falls on these gratings, different

wavelength components are reflected by the grating array at different times according to

the arrangement of the grating elements causing the wavelength components of the light

pulse to be scrambled in time. The receiver restores the original signal using the same

setup with an array of gratings that has the reverse pattern to that of the transmitter

side.

2.3.3   Hybrid Wavelength-Time Spreading

In this method the same setup used in Section 2.3 can be used with a modification to

define the distances between the different grating cells in the FBG according to pseudo-

random spreading code [35]. This produces time spreading between the different wavelength components as well as the spectral spreading resulting from scrambling the wave-lengths. This spreading method requires a 2-D spreading code [60], which uses one dimension for wavelength spreading and the other dimension for time spreading.

2.4 Optical Spreading Codes (OSC)

Since optical transmission is based on carrying the information using the signal power, many of the coding methods used in electrical-based CDMA systems (such as wireless RF systems) are not suitable for OCDMA since their properties are based on the assumption that signal amplitudes can be considered positive or negative (phase change is used). This is not practical in optical transmission systems. Hence, spreading codes such as Walsh codes, maximal length sequences, and Gold sequences, which have desired properties suitable for wireless CDMA, are not usable in optical transmission.

New families of spreading codes that are suitable for optical signals have been investigated in the literature. The most common two families of optical spreading codes are: Optical Prime Codes (OPC) (see [61, 62] and citations within) and Optical Orthogonal Codes (OOC) (see [63, 64] and citations within).

2.4.1 Optical Prime Codes

Optical Prime Codes (OPC) are designed using linear congruence theory [58]. A prime codeword C_j from the family C of p codewords constructed over the Galois field $(GF(p))$

of a prime number p is given by:

$$C_j = \{c_{ij} \; : \; c_{ij} = (i \cdot j \mod p), \quad i,j \in GF(p)\}\,.$$

Prime codes have codewords with code length $n = p^2$ and code weight $k = p$. They have a crosscorrelation value that is at most equal to 2 and high autocorrelation value.

2.4.2 Optical Orthogonal Codes

In Optical Orthogonal Codes (OOC) [65], the circular[2] autocorrelation with time shift $\tau \neq 0$ for a codeword and the circular crosscorrelation between any two code words are minimized (or limited) while the zero-shift circular autocorrelation for a codeword is maximized.

In general, and throughout this thesis, when we refer to Optical Orthogonal Code (OOC) we are referring to the case where the non-zero shift autocorrelation and the crosscorrelation are limited by a maximum value of 1. These codes posses the unique property that there is a maximum of a single '1'-chip (pulse) overlap between any two different codewords belonging to the same family with arbitrary circular time shifts and a maximum of a single '1'-chip overlap between a code and a circular time shifted version of itself.

Detailed definition and construction methods for OOCs are provided in Chapter 6.

[2]Circular here refers to the fact that the codewords are correlated using a circular bit rotation instead of a shift in one end direction with zero insertions from the other end

2.5 OCDMA Decoding Methods

Techniques used for decoding OCDMA signals at the receiver side have a significant impact on the overall optical system performance and complexity. In the literature, two main techniques are used to decode OCDMA signals; namely, correlation detection and chip-level detection.

2.5.1 Correlation Detection

In correlation detection [66, 67], decoding the signal is performed by counting the total number of photons at marked chip positions (i.e., chip positions for 1-chips according to the spreading code) over the entire bit duration. Determining marked chip positions is done either using a passive decoder composed of fiber tapped delay lines followed by a timed optical gate (to open for one chip duration at the end of the bit duration), or an active decoder using a timed optical source that only produces pulses at the marked chip positions to be correlated with the received signal [67]. The total count is compared to a certain threshold. If it is higher than the threshold it is considered a '1' bit, otherwise it is a '0' bit.

2.5.2 Chip-Level Detection

Chip-level detection [68, 69] is performed by counting the photons over every marked chip individually and comparing that number with a threshold. If the number in all marked chips is higher than the threshold a '1' bit is detected, otherwise it is a '0' bit.

Each technique has its advantages and disadvantages. While the chip-level detection is less complex in implementation, it is a technique that cannot be performed using optical passive devices and it has lower performance [68] relative to correlation detection. In contrast, correlation detection achieves higher performance and can be implemented using passive optical devices but it requires higher system complexity [68].

2.6 Encoding/Decoding of Time Spread OCDMA

The principles of time domain spreading of optical signals using optical correlation encoding and decoding are illustrated in Figure 2.4. As shown in the figure, the ODLC based encoder takes as an input a very short optical pulse with duration T_c that represents either a logical '1' or a logical '0'. The pulse then goes through several parallel tapped delay lines using an optical splitter, thus experiencing different delays over these lines. The pattern of the delay across the delay lines specifies the spreading code. The delay values for the delay lines of the ODLC encoder are chosen to produce an encoded signal that is spread using Optical Orthogonal Codes (OOC). The outputs of these delay lines are combined together using an optical coupler and then amplified to overcome the attenuation caused by both the splitter and the coupler. The ODLC based decoder has the same structure as the encoder with the delay pattern chosen so that they can reverse the encoder effect and restore the original intended signal from a mix of multiplexed signals using different orthogonal codes. This is achieved using a matched filter design at the decoder side as will be explained later in this section.

The receiver has an associated *threshold detector* (see Figure 2.4), which only passes light if its total energy is above a certain value. If we assume that the energy per spread chip pulse is $\mathcal{Q}$, then the total energy of the de-spread pulse is $k \times \mathcal{Q}$, where k is the spreading code weight. The threshold device will pass light only if its total energy is greater than or equal to $h \times \mathcal{Q}$, where $0 \leq h \leq k$ is the energy threshold factor, k is the code weight of the spreading code and its code length is n. The use of the threshold detection device ensures that the low amplitude pulses that are uncorrelated with the decoder delay pattern are blocked from causing background noise to the correlated pulse. The device also helps against dark current effects, which are inherent in any optical detection system [70].

To complete the functional specification of the ODLC unit, it remains to show how to select the delay patterns at the transmitter OOC encoder to realize the required OOC spreading and at the receiver OOC decoder to recover (despread) the original signal.

Lemma 2.1. *Let time be measured in increments of T_c and let the input to the ODLC encoder be given by the optical chip pulse $p(t)$ with duration T_c. Assuming that the ODLC encoder is a Linear Time Invariant (LTI) system, the impulse response $h_i(t)$ for an ideal ODLC encoder used for encoding $p(t)$ into encoded signal $y_i(t)$ using code number i, is given by:*

$$h_i(t) = \frac{1}{k} \sum_{m=1}^{k} \delta(t - e_m^{(i)} T_c) \, , \tag{2.1}$$

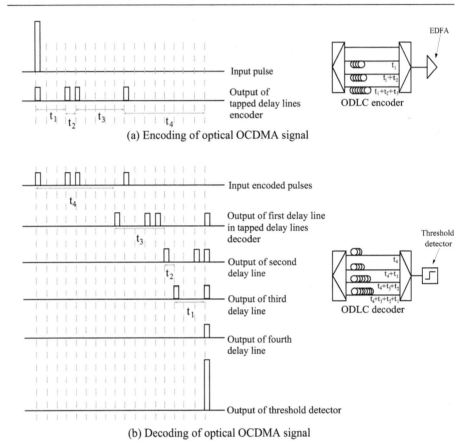

(a) Encoding of optical OCDMA signal

(b) Decoding of optical OCDMA signal

Figure 2.4: Encoding and decoding of time spread fiber optic OCDMA signals.

where

$$
e_m^{(i)} \equiv
\begin{cases}
0 & m = 1 \\[2ex]
\displaystyle\sum_{j=1}^{m-1} t_{i,j} & 2 \le m \le k
\end{cases}
,
\tag{2.2}
$$

and $\delta(\cdot)$ is the Dirac-Delta function. The desired encoded output $y_i(t)$ is defined as:

$$y_i(t) \equiv \frac{1}{k} \sum_{m=1}^{k} p(t - e_m^{(i)} T_c) \tag{2.3}$$

Proof. Since the encoder is assumed to be a LTI system, we can write:

$$y_i(t) = p(t) * h_i(t) . \tag{2.4}$$

Rewriting (2.3) as

$$y_i(t) = \frac{1}{k} \sum_{m=1}^{k} p(t) * \delta(t - e_m^{(i)} T_c) ,$$

which can be written as

$$y_i(t) = p(t) * \frac{1}{k} \sum_{m=1}^{k} \delta(t - e_m^{(i)} T_c) .$$

Comparing this with (2.4) we get

$$h_i(t) = \frac{1}{k} \sum_{m=1}^{k} \delta(t - e_m^{(i)} T_c) .$$

□

Theorem 2.2. *Let time be measured in increments of T_c and let the input to the ODLC decoder be given by $y_i(t)$ defined in (2.3). Assuming that the ODLC decoder is a Linear Time Invariant (LTI) system, the impulse response $q_i(t)$ for an ideal matched filter ODLC decoder used for decoding $y_i(t)$ into the original transmitted chip pulse $(p(t))$ is given by:*

$$q_i(t) = \sum_{m=1}^{k} \delta(t - d_m^{(i)} T_c) \ , \tag{2.5}$$

where

$$d_m^{(i)} = \sum_{j=m}^{k} t_{i,j} \ . \tag{2.6}$$

Proof. A matched filter to $h_i(t)$, deduced in Lemma 2.1, is given by

$$
\begin{aligned}
h_i^*(nT_c - t) &= \frac{1}{k} \sum_{m=1}^{k} \delta(nT_c - t - e_m^{(i)} T_c) \\
&= \frac{1}{k} \sum_{m=1}^{k} \delta(-t + [n - e_m^{(i)}] T_c) \\
&= \frac{1}{k} \sum_{m=1}^{k} \delta(t - [n - e_m^{(i)}] T_c) \\
&= \frac{1}{k} \sum_{m=1}^{k} \delta(t - d_m^{(i)} T_c) \ ,
\end{aligned}
$$

where $d_m^{(i)} \equiv n - e_m^{(i)}$.

Our next step is to find an expression for $d_m^{(i)}$. Using (2.2), we can write:

$$d_m^{(i)} = n - e_m^{(i)} = \sum_{j=1}^{k} t_{i,j} - \begin{cases} 0 & m = 1 \\ \sum_{j=1}^{m-1} t_{i,j} & 2 \le m \le k \end{cases}$$

$$= \begin{cases} \sum_{j=1}^{k} t_{i,j} & m = 1 \\ \sum_{j=1}^{k} t_{i,j} - \sum_{j=1}^{m-1} t_{i,j} & 2 \le m \le k \end{cases}$$

$$= \begin{cases} \sum_{j=1}^{k} t_{i,j} & m = 1 \\ \sum_{j=m}^{k} t_{i,j} & 2 \le m \le k \end{cases}$$

$$= \sum_{j=m}^{k} t_{i,j} \ .$$

Finally, we find the correct amplitude scaling factor so that the decoder can retrieve the original signal $p(t)$. Since $h_i(t)$ has a factor of $\frac{1}{k}$, the decoder should counter this effect by having a factor k multiplied by the response of the filter matched to the transmitter. Therefore,

$$q_i(t) = kh_i^*(nT_c - t) = \sum_{m=1}^{k} \delta(t - d_m^{(i)}T_c) \ .$$

$\square$

Equations (2.1) and (2.2) describe the design of an ideal ODLC encoder, while (2.5) and (2.6) describe an ideal ODLC decoder. When applying these equations to the design of ODLC encoders and decoders, we find that an encoder for code i with k delay lines will have the delay value $e_j^{(i)} T_c$ at optical delay line $j \in \{1, 2, ..., k\}$. While $d_j^{(i)} T_c$ is the delay value at optical delay line j of the ODLC decoder for code i. If the components of the encoder/decoder are not ideal (i.e., there are fiber losses proportional to length or asymmetry in the splitters/combiners) appropriate optical amplification and splitting/combining ratios can be used to design the ODLC so that the same equations can be obtained again.

Example 2.3. *To illustrate the use of the set of inter-delays representation in designing the ODLC, consider the encoding/decoding example shown in Figure 2.4. Let $k = 4$ and $N = 1$ (a trivial code matrix is used for illustration purposes), then the set*

$$\mathbb{T} = \{t_{11}, t_{12}, t_{13}, t_{14}\}$$

is the set of inter-delays between the symbols. Forming the extended set of inter-delays $\mathbb{T}_{ext}$ by taking into consideration the sum of l adjacent delays up to $l = k - 1$ for every delay element in $\mathbb{T}$, one gets

$$\mathbb{T}_{ext} = \{t_{11}, t_{12}, t_{13}, t_{14}, t_{11} + t_{12}, t_{12} + t_{13}, t_{13} + t_{14},$$
$$t_{14} + t_{11}, t_{11} + t_{12} + t_{13}, t_{12} + t_{13} + t_{14},$$
$$t_{13} + t_{14} + t_{11}, t_{14} + t_{11} + t_{12}\} ,$$

which must have unique elements.

Referring to Figure 2.4-(a), the encoder delays in steps of chip duration T_c are $\{0, t_{11}, t_{11} +$

$t_{12}, t_{11} + t_{12} + t_{13}\}$. At the receiver, to de-spread the signal the decoder delays in steps of

T_c must be $\{t_{14}, t_{14} + t_{13}, t_{14} + t_{13} + t_{12}, t_{14} + t_{13} + t_{12} + t_{11}\}$ as shown in Figure 2.4-(b).

It can be seen from Theorem 2.2 and from the example above that the decoding

ODLC produces k shifted versions of the received sequence such that at the output of

each delay tap the location of one of the k pulses of the spreading code coincides with

the chip location right after the last chip of the bit duration T_b. Accordingly, if at the

output there is an optical timed gate (a sampling device) that only opens at this chip

location allowing light to pass through an optical threshold device and to be detected by

a photodetector, it is effectively as if this receiver is counting the sum of photons over

chip locations corresponding to pulse locations in the spreading code and comparing that

with a threshold value. This is exactly the function described earlier for the correlator

receiver.

2.7 Concluding Remarks

We have provided a brief introduction to the subject of using code division multiple

access in optical fiber communication systems. In this introduction, we have discussed

the transceiver architecture of OCDMA systems. We have provided a brief description of

the digital modulation methods commonly used in OCDMA systems while describing the

cons and pros of each method. We followed by discussing the methods used for spreading optical signal in time and wavelength domains. Two major families of optical spreading codes were introduced and their properties as well as primary construction methods were explained. The main categories of optical detection methods were discussed explaining briefly the advantages and disadvantages of each.

The chapter also provided a detailed description for a class of OCDMA encoder/decoder devices called Optical Delay Lines Correlators (ODLC), which is used for optical time spreading. The architecture of ODLC using all-optical components as well as its operation as a spreading device were explained in detail. Following that, a mathematically derived design of these devices was provided. Using the mathematical design model, it was shown that these devices belong to the category of optical decoders called *correlator detectors*.

In the next chapter, we will propose a modification to the architecture of Generalized Multi Protocol Label Switching GMPLS in order to utilize OCDMA as a switching mechanism in all-optical (photonic) switching networks.

Chapter 3

OC-GMPLS Core Sub-Wavelength

Switching Architecture

In Chapter 2, an overview of the technologies and methodologies that enable the use of OCDMA has been provided. In this chapter, we propose the use of OCDMA to expand the available label space in GMPLS based all-optical networks. Our proposal provides an efficient solution for finer granularity *"sub-wavelength"* all-optical switching in the core of the network, which significantly enhances the utilization and the traffic isolation capability of optical GMPLS networks.

The rest of this chapter is organized as follows. We start by a brief introduction in Section 3.1. We introduce the code switch capable layer as a new label switching layer for GMPLS networks in Section 3.2. Section 3.3 describes the end-to-end architecture of the Optical Code Labeled GMPLS (OC-GMPLS) network and the label stacking mechanism. In Section 3.4, we propose a high level architecture for a core OC-GMPLS switch. Following that, we introduce the architecture for the edge OC-GMPLS switch in Section 3.5. The performance of the proposed core switch architecture in terms of the number of isolated traffic flows that can be switched at the network core is analyzed in

Section 3.6. Numerical results for our analytical model are discussed in Section 3.7. We conclude the chapter in Section 3.8.

3.1 Introduction

Code Division Multiple Access (CDMA) [71] is a physical layer multiplexing technique that can be utilized in optical networks as a means for switching several distinct data flows over the same fiber and the same wavelength entirely in the all-optical domain. The concept of using optical code as a labeling mechanism in MPLS systems has been suggested before in the literature [36, 72–74]. Labeling in the case of MPLS is performed by using the optical code as a method to encode addresses into the packet header only. This is what is referred to as *Explicit Optical Code Labeling* (ExOCL) method [25]. Another application of ExOCL is the use of CDMA techniques in Optical Burst Switching (OBS) [75, 76].

The current standard for GMPLS [77] defines five *label mapping spaces* namely: Packet Switch Capable (PSC), Layer 2 Switch Capable (L2SC), Time-slot Switch Capable (TSC), Wavelength Switch Capable (LSC), and Fiber Switch Capable (FSC). Only the last two layers (LSC and FSC) can be utilized in an all-optical switching device. The remaining label mapping spaces require optical-to-electrical conversion, which negatively impacts the switching speed.

We propose a new GMPLS label mapping layer called the Code Switch Capable (CSC) layer, which can perform labeling and switching entirely in the optical domain using optical spread codes. It should be noted that the TSC, L2SC, and PSC layers provide sub-wavelength switching in GMPLS networks. However, they suffer from the serious limitation of requiring optical-to-electrical-to-optical conversion. This means that these switching layers are more suitable for optical edge switching devices, where a higher level of traffic isolation is more important than switching speed and capacity.

The proposed CSC layer provides the main capability for achieving all-optical sub-wavelength switching in GMPLS core networks. A capability which proves useful in increasing the granularity of traffic isolation in the network core, where all-optical switching is needed the most.

Enhancing the GMPLS architecture (by allowing the use of the new label mapping space) will have significant impact on advancing the QoS and traffic engineering capabilities of optical networking especially at the network core. Specifically, providing a seamless combination (or cross product) of code-space and wavelength-space into a larger single label space increases flow granularity at the all-optical switching level, and hence, the utilization in all-optical networks without impacting the complexity of the optical control plane.

3.2 Code Switch Capable (CSC) Layer

We propose to modify the GMPLS label mapping methods by adding the Code Switch Capable (CSC) layer (or interface), which can perform labeling and switching based on optical spreading codes. We call this version of GMPLS Optical Code Labelled GMPLS

(OC-GMPLS). Figure 3.1 shows the label mapping space in OC-GMPLS, and how the CSC layer is added in between the TSC layer and the LSC layer. In the following, we define the CSC layer in a manner similar to what is done for the other existing label mapping layers in GMPLS.

Definition 3.1. *The CSC interface is capable of forwarding data based on an optical code that is used to spread the data packet.*

The CSC layer is placed at a level between the TSC layer and the LSC layer due to the following reasons:

- The CSC interface switching functionality can be performed entirely in the optical domain. This fits with all the layers below it; namely the LSC and FSC layers. On the other hand, all layers above the CSC, viz. TSC, L2SC, and PSC, require optical to electrical conversion in order to perform switching functions.

- The number of flows that can be supported by the CSC interface is higher than that supported by the LSC (wavelength) layer, but lower than the number of flows that can be supported by the TSC layer.

In order to integrate this scheme into the GMPLS framework, we utilize the *Implicit Optical Code Labeling* (ImOCL) method for deploying OCDMA as a labeling mechanism in GMPLS. In ImOCL the optical spread code is applied to the whole data stream instead of attaching an OCDMA encoded header to the packet as in the case of the

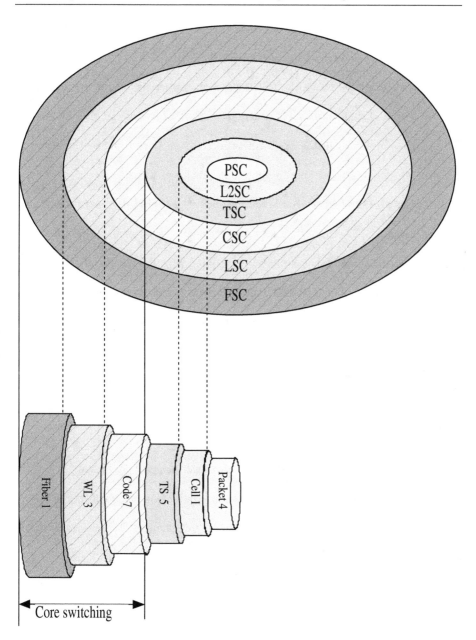

Figure 3.1: Label mapping layers in OC-GMPLS.

ExOCL method [25]. The ImOCL method is simple to implement and fits within the physical layer labeling mechanisms used in GMPLS.

The purpose of introducing the CSC layer is to further extend the benefits of the label mapping technique in the all-optical domain. In our case one can draw an analogy between the wavelength in the optical network and the shared link in the electrical network. Our goal is to be able to apply different labels to packets carried over the same wavelength (λ) while being able to label, detect, and filter packets entirely in the all-optical domain. This means that using our technique, even in an all-optical (photonic) network, the bandwidth granularity is no longer limited by the number of wavelengths, but rather by the number of optically labelled data within the wavelengths. This, in our view, is useful for the deployment of next generation optical switching nodes with sub-wavelength switching capabilities at the network core. Label space expansion with CSC can be illustrated by considering an example of an optical λ switch, which supports $M = 20$ wavelengths. If the CSC interface can assign up to $N = 50$ labels to optical data, this creates $M \times N = 1000$ distinct optical flows, which can be switched entirely in the optical domain without any optical-to-electrical-to-optical conversion.

In order to utilize the capabilities provided by the CSC layer, modifications are required to existing GMPLS switch architectures as well as overall network architecture and signalling. These modifications will be the focus of the rest of this chapter.

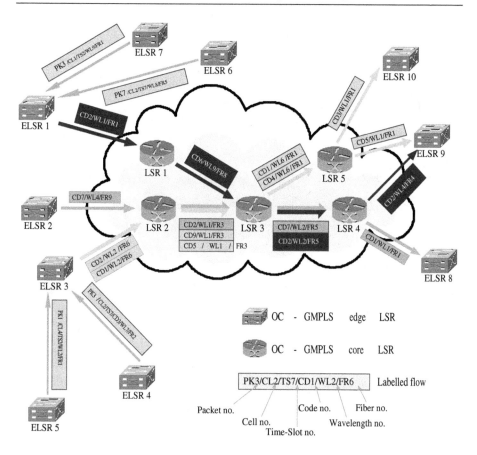

Figure 3.2: OC-GMPLS network architecture and end-to-end operation.

3.3 End-to-End Network Architecture

We propose a network architecture for the OC-GMPLS framework that utilizes the hier-archical topology using the Core-Edge concept [78]. Our proposal utilizes only three of the OC-GMPLS switching layers in the network core. These layers are the FSC, LSC, and CSC. The reason for choosing these layers at the network core stems from the fact that these are the only three layers that can perform switching in an all-optical manner. This allows for an entirely all-optical core, which is capable of reasonable traffic isolation at the sub-wavelength level using OCDMA, while maintaining the high speed and high capacity features of all-optical switching. At the network edge, however, all the label mapping and switching layers of the OC-GMPLS architecture are utilized in order to allow for the highest possible level of traffic isolation. It is possible for the network operators to choose not to use the CSC layer in case of edge to edge traffic as in the case of traffic between ELSR 5 and ELSR 3 in Figure 3.2 or to use it as in the case of traffic between ELSR 4 and ELSR 5. The reason for allowing this option is that with the existence of TSC, L2SC and PSC switching capability the CSC does not expand the label mapping space. In fact these three label mapping layers result in much larger label mapping space alone. However, the capability itself has to exist in the switch in order to allow for exploiting its benefits at the core switching level.

The edge-to-edge network architecture as well as an example label stacking scenario is shown in Figure 3.2. It can be seen from the figure that the core labelled traffic is identified only by three levels of label stacks; namely the fiber number, the wavelength number and the optical code number. For edge traffic, however, labeling is performed using up to six levels of labels, which are the time-slot number, the cell number and the packet number in addition to the three mentioned in the core. It is the function of the edge devices, such as ELSR 1, to perform aggregation of higher label mapping layers into a single labelled traffic in the core of the network. This process treats the aggregated traffic as a single flow at the core level. Using the CSC capability, the edge device can perform aggregation into core traffic using different codes, which enables traffic isolation at the core level at the sub-wavelength level as shown by the aggregation process performed by ELSR 3 using two different codes on the same wavelength and fiber. The aggregation process can be performed using our proposed Containerization with Aggregation Timeout (CAT) aggregation method [79].

Exploiting the same capability (i.e., switching at the sub-wavelength optical code level) by the core routers, leads to a significant increase in the all-optical core switching capacity of optical GMPLS networks as will be illustrated later in this chapter.

In order to allow signalling for path setup operations and other network control operations, a dedicated CSC label or group of labels are used for signalling and control operations. These are shown in Figure 3.3 by the lighter grey color arrows at the input and output of the OXC unit.

3.4   Core Switch Architecture

Implementing OC-GMPLS, with the new CSC layer, requires certain modifications to current optical switch architectures. The key new capability introduced is optical code processing of data packets to realize the spreading and de-spreading functions necessary for OCDMA. Figure 3.3 shows the model of the OC-GMPLS core switch, which is capable of performing optical label switching using the first three all-optical switching layers

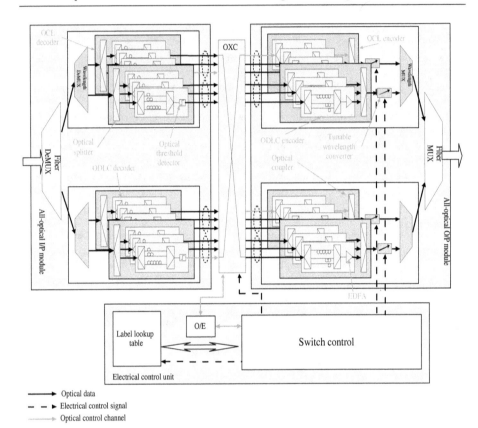

Figure 3.3: Architecture of OC-GMPLS sub-wavelength core switch.

(FSC, LSC, and CSC) of the GMPLS label mapping model.

The switch consists of four main parts; the Optical Cross-Connect (OXC) backplane, the all-optical input module, the all-optical output module and the electrical control unit. The all-optical input module consists of a fiber de-multiplexer that allows for separating incoming fibers into different ports of the switch. The output of each single fiber is fed into a wavelength de-multiplexer, which separates different wavelengths of the input

Dense Wavelength Division Multiplexed (DWDM) signal. Every wavelength is split using

an optical splitter through a rack of N different Optical Delay Line Correlator (ODLC)

decoders each of which is matched to one of the N different optical codes. This rack

of ODLC decoders combined with the optical splitter at the input is called the Optical

Code Labeling (OCL) decoding unit. To make the switch non-blocking internally, $M \times L$

copies of OCL decoders are used, where M is the number of wavelengths and L is the

number of fibers.

The all-optical output module performs the reverse of the above operation. It starts

with an OCL encoding unit, which has the same structure of the OCL decoding unit

except that it consists of N ODLC encoders connected to an optical coupler. Again $M \times L$

copies of the OCL encoders are needed to guarantee that the switch is internally non-

blocking. The outputs of the OCL encoders are fed into tunable wavelength converters,

which modulate it on a wavelength selected by the control unit. The outputs of the

different wavelength converters are combined using a wavelength multiplexing unit to

compose the DWDM signal for each fiber. Finally, the DWDM signals are distributed

into the different fibers using the fiber multiplexer.

The electrical control unit in conjunction with the core OXC function as the heart of

the switch. The control unit performs the label switching operation by making sure that

the OXC is configured in a manner that it switches incoming flows belonging to a certain

Label Switched Path (LSP) to their corresponding output Forward Equivalent Class

(FEC). The control unit also maintains the label lookup table. The control unit receives and sends necessary signalling and control data for GMPLS network configuration and maintenance using the dedicated signalling and control labels.

The function of the ODLC unit is to perform optical correlation of the pulse train with the time transfer function of the tapped delay lines system to encode data pulses into spread pulse trains, or to decode spread data pulse trains into data pulses. Each ODLC has a unique pattern of k optical delay lines that defines the spreading/de-spreading code. The function of the ODLC as well as the principles of optical correlation encoding and decoding used to design the ODLC encoders/decoders are explained in Chapter 2. For the purpose of this chapter, it suffices to assume that the output of the ODLC encoder is a train of pulses with a pattern that produces a time encoded Optical Orthogonal Code (OOC) signal with length n and code weight k. The output of the decoder is the restored original signal before encoding.

3.5 Edge Switch Architecture

The edge switch shown in Figure 3.4 consists of two main units. The first is the optical core part, which performs switching of the three core layers (FSC, LSC, and CSC). It has exactly the same design and functionality as the core switch described previously. The second unit is added for the upper three layers (TSC, L2SC, and PSC) switching. This unit operates in the electrical domain. This is why an optical to electrical conversion is needed before inserting data into this unit, then an electrical to optical conversion is

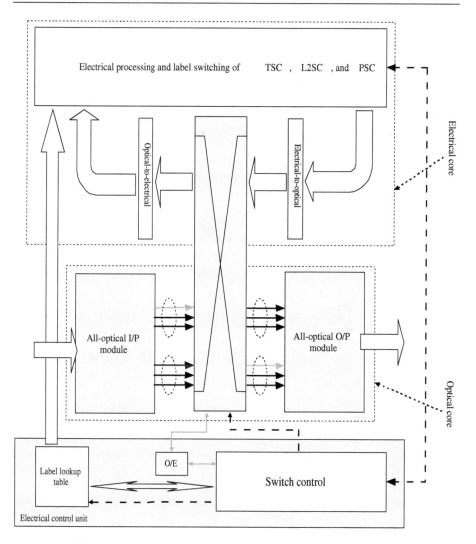

Figure 3.4: Architecture of OC-GMPLS sub-wavelength edge switch.

needed before inserting the output of this part into the optical core part. The connection between the two units is established through the OXC unit, which is expanded to allow all data coming through the optical interfaces to be fed into the electrical processing unit without any internal blocking. The control unit as well as the label lookup table are also extended to support the new functions.

The advantage of the proposed edge switch design is that it is simplified by making part of its structure common with the core switch and in this part the switching operation can still be performed in an all-optical mechanism.

3.6 Switching Capacity and Label Space Expansion Ratio

The performance of our proposed system will be analyzed in this section from the aspect of the effect of the proposed architecture on the all-optical core label space mapping (or the flow switching capacity) as compared to a system that uses FSC and LSC only. This will be represented by a factor called the Label Space Expansion Ratio ρ.

Assuming that the energy per spread pulse at the output of the ODLC encoder (or equivalently the input of the ODLC decoder) is Q, then the total energy of the de-spread pulse is $k \times Q$. The threshold device will pass light only if its total energy is greater than or equal to $h \times Q$, where $0 \leq h \leq k$ so that the desired de-spread pulse can be detected. Note, however, that there is always a possibility of detection errors if the number of orthogonal codewords (N) is greater than the detection optical threshold h. Errors can happen when h or more spread pulses from h or more flows other than the intended flow occur during the same chip time T at the output of the decoder correlator. In this case, the energy of the pulse resulting from the addition of the spread pulses will exceed the threshold value and will be incorrectly detected as an intended received pulse. This suggests that an error free receiver system for OOC CDMA can be achieved if the condition

$$N \leq h \leq k \tag{3.1}$$

is satisfied, i.e., the threshold parameter h must be smaller than k (the code weight) but larger than N (the number of orthogonal codewords or labels within the same wavelength).

The circular autocorrelation and crosscorrelation properties for OOC spreading codes are summarized as [25]:

$$\mathcal{R}_{XX}(\tau) = \begin{cases} k & \tau = 0 \\ \leq 1 & 1 \leq \tau \leq n-1 \end{cases}$$

$$\mathcal{R}_{XY}(\tau) \leq 1, \qquad 0 \leq \tau \leq n-1,$$

where $\mathcal{R}_{XX}(\tau)$ is the circular autocorrelation function for sequence X and $\mathcal{R}_{XY}(\tau)$ is the circular crosscorrelation function between sequences X and Y. A more formal definition for OOC and their autocorrelation and crosscorrelation will be given in Chapter 6.

Using set theory it was shown in [15,30] that the autocorrelation properties correspond

to the condition that for a *non-trivial OOC*[3] to exist, we must have

$$k(k-1) \leq n-1 \, , \tag{3.2}$$

where n is the code length. Moreover, the number of orthogonal codewords N, in a

non-trivial OOC, is limited by

$$N \leq \left\lfloor \frac{n-1}{k(k-1)} \right\rfloor \, . \tag{3.3}$$

For a DWDM system that uses M wavelengths and L fibers with a typical channel

bandwidth of W Hz (typically $W = 100$ GHz), the maximum number of optically isolated

or identifiable flows (F_λ) is given by:

$$F_\lambda = L \times M \, . \tag{3.4}$$

In order to increase this number, we have to perform aggregation of signals in the electrical

domain and convert the aggregate signal to the optical domain. This means that we will

need faster electronics to carry out the operation and, at the same time, we will loose

flow identification at the optical level in intermediate nodes.

On the other hand, assuming that the bandwidth of a flow is given by B, where

---

[3]A non-trivial OOC is a code that satisfies the condition: $N > 1$.

typically $B << W$ (e.g., B can be in the order of 100 MHz - 1 GHz), we can use a code

of length $n = \lfloor \frac{W}{B} \rfloor$ to multiplex N optically isolated flows on a single wavelength, where

N is given by (3.3). Then, the maximum number of flows (F_C) that this new system can

carry is given by:

$$F_C = L \times M \times N \leq L \times M \times \left\lfloor \frac{n-1}{k(k-1)} \right\rfloor . \qquad (3.5)$$

The maximum value for the maximum number of flows (F_C^*) in this case will hold in the

equality case. In other words, F_C^* is given by:

$$F_C^* = L \times M \times \left\lfloor \frac{n-1}{k(k-1)} \right\rfloor . \qquad (3.6)$$

Defining the label space expansion ratio (ρ) to be the ratio of the number of labels using

OCDMA for sub-wavelength labeling (i.e., based on a combination of OCDMA and λ in

addition to fiber) to the number of labels using DWDM (i.e., based on λ in addition to

fiber), we get

$$\rho = \frac{F_C}{F_\lambda} . \qquad (3.7)$$

Substituting (3.4) and (3.5) into (3.7) gives

$$\rho = N \leq \left\lfloor \frac{n-1}{k(k-1)} \right\rfloor . \qquad (3.8)$$

Consequently, the maximum label space expansion ratio ρ^* is achieved when the equality

holds and is given by:

$$\rho^* = \left\lfloor \frac{n-1}{k(k-1)} \right\rfloor .$$
(3.9)

For an error free transmission, we have

$$\rho_{nerr} = N \leq h \leq k ,$$
(3.10)

where ρ_{nerr} is the value of ρ for an error free system.

In order to maximize the label expansion ratio and minimize the MAI error rate, we

choose [25]

$$h = k$$
(3.11)

Using (3.8), (3.10) and (3.11) we can find the approximate maximum error free label

expansion ratio ρ^*_{nerr} by formalizing a continuous approximation to the discrete optimiza-

tion problem and rounding the solution to the lowest integer. The continuous counterpart of the optimization problem is given by:

$$\max_{k} \quad \rho$$

$$\text{subject to} \quad \rho \leq \left\lfloor \frac{n-1}{k(k-1)} \right\rfloor$$

$$\text{and} \quad \rho \leq k \, ,$$

where we consider all the variables k, ρ, and n to be continuous when searching for an optimum solution. The Lagrange multiplier method can be used to formally solve this optimization problem and deduce ρ^*_{nerr}. However, the solution can be deduced directly from the constraints and the cost function or graphically as shown in Figure 3.6. We give the final results in (3.12) and (3.13).

$$k^*_{nerr} = \left\lfloor \frac{n-1}{k^*_{nerr}(k^*_{nerr}-1)} \right\rfloor \tag{3.12}$$

$$\rho^*_{nerr} = \left\lfloor \frac{n-1}{\rho^*_{nerr}(\rho^*_{nerr}-1)} \right\rfloor , \tag{3.13}$$

where ρ^*_{nerr} is the optimum (maximum value) for label space expansion ratio under error free multiplexing conditions and k^*_{nerr} is the code weight value that achieves the optimum value for ρ.

For an efficient OCDMA system, the values of ρ and n must satisfy $\rho \gg 1$ and

$n \gg 1$. Accordingly, (3.13) can be approximated by:

$$\rho^*_{nerr} \approx \lfloor \sqrt[3]{n} \rfloor \ . \tag{3.14}$$

It can be seen from (3.13) that the optimum value for the label space expansion ratio in an error free system is independent of the code weight, which means that choosing the value for the code length n determines the value of the maximum error free label space expansion ratio.

3.7   Numerical Results

In our numerical results, we use a code length $n = 1000$. This value is close to actual practical values as the channel in a DWDM system has a bandwidth of about 100 Gbps for a binary transmission scheme and a standard traffic source such as an ethernet work-station produces data rates of about 100 Mbps, which gives a ratio $n = \lfloor \frac{W}{B} \rfloor = 1000$. We also choose the number of wavelength channels to be $M = 50$, which complies with the ITU-T DWDM standards [80].

In Figure 3.5, the different regions of system operation are shown. There are two main regions considered in this system. The first is the region under which the orthogonality condition is maintained, which is defined by the inequality in (3.3). This is marked by the line labeled N in the figure. The other important region is the region in which the system operates with no MAI errors. This latter region is defined by the inequality $N \leq h$ and is marked by the shaded area in Figure 3.5. In the same figure we show how the system can

be loaded with a certain number of users while maintaining the orthogonality conditions

under different values for the code weight k. This is shown by the thick line. It is worth

mentioning here that for an OCDMA system to operate without errors, the conditions in

(3.10) have also to be fulfilled. A system that fulfills all the three conditions is illustrated

in Figure 3.6.

Figure 3.6 shows a plot of the label space expansion ratio ρ versus the code weight

k based on (3.8) and (3.10) for a value of $n = 901^4$. The shaded area in the figure

represents the range of ρ over which the system is working under error free conditions

with the orthogonality conditions maintained. The figure demonstrates the existence of

an optimum value for ρ, which can be derived by graphically finding the intersection

point that provides the solution to (3.8) and (3.10) simultaneously. For example, for

$n = 901$, the optimum code length is $k^*_{nerr} = 10$ and the optimum label expansion ratio is

$\rho^*_{nerr} = 10$, which means that our method can generate ten times the labels of a DWDM

system without affecting the error performance of the system.

The maximum value of k depends on the number of orthogonal codes (desired max-

imum number of flows) N as indicated in (3.3). Figure 3.6 verifies that increasing k

reduces the label space expansion ratio ρ because of the decrease in the number of codes

N to fulfill the orthogonality condition. This was expressed mathematically in (3.8).

However, it must be noticed that the optimum label expansion value depends solely on

---

^{4}This value is selected to give an integer value for $\frac{n-1}{k(k-1)}$.

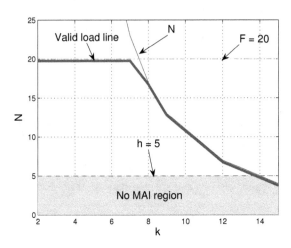

Figure 3.5: Operational regions and load lines for OCDMA systems.

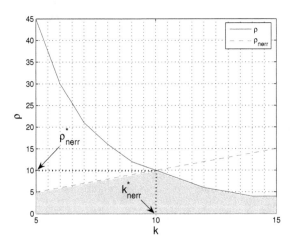

Figure 3.6: Finding the optimum label space expansion ratio for code length = 901.

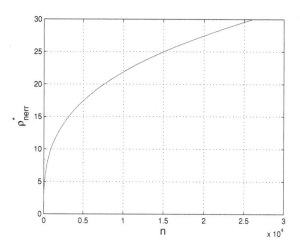

Figure 3.7: Optimum label space expansion ratio.

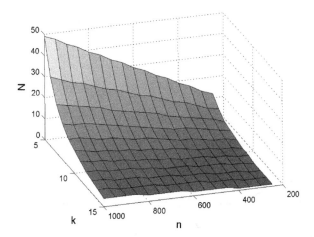

Figure 3.8: Effect of code weight and code length on the number of orthogonal codes.

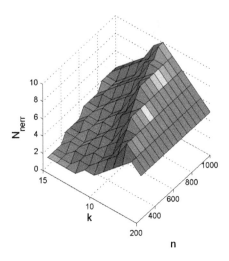

Figure 3.9: Effect of code weight and code length on the number of orthogonal codes in the error free region.

the code length n as shown in (3.13) and demonstrated by Figure 3.7.

Figure 3.8 shows the effect of n and k on the maximum number of possible orthogonal codes N. It can be seen from the figure that k has a more prominent effect on N than n, which suggests that tuning k to the optimum value is more important.

In Figure 3.9, we show the same thing with the condition that the system is operating under no MAI error. It can be seen from the figure that there is a maximum value for N_{nerr} that occurs at a certain value $k = k_{nerr}$ for each value of n. It can also be seen that increasing n increases the maximum (optimum) value of N_{nerr}. Figure 3.9 can be thought of as multiples of Figure 3.6 plotted at different values of n, which makes it easier to interpret the figure.

3.8 Concluding Remarks

In this chapter, we have proposed an extension to the GMPLS framework that increases its scalability by expanding the label mapping space at the network core. Increasing the label space in GMPLS is a very important capability since it enables a finer granularity through sub-wavelength switching to be achieved entirely in the all-optical domain. The expansion to the label mapping space was achieved by exploiting a physical layer technique based on optical CDMA.

An end-to-end system architecture that involves the hierarchical edge-core architecture is proposed in conjunction with an architecture for both edge switches and core switches. We then propose an implementation of the architecture using time domain spreading with fiber tapped delay line correlators and optical orthogonal codes for spreading.

Using basic elements of set theory, we have provided a mathematical analysis for the proposed technique, which demonstrates that our expansion method of the core label mapping space is equal to a factor of ρ multiplied by the number of available wavelength channels multiplied by the number of fibers. This factor depends on the code design used for optical CDMA and introduces design tradeoffs to achieve the maximum expansion while maintaining the system operation error free.

Using analytical methods, as well as numerical evaluations, we were able to derive the system optimal operating point in order to maximize the label space expansion ratio (sub-wavelength switching capacity) while maintaining the system operation in the error free MAI region. Our results show that the value of the optimum label space expansion ratio is dependent only on the length of the OCDMA spreading code. This means that for a system that limits this value, due to bandwidth restrictions, the limits on the label expansion ratio can be directly deduced. It was also shown that achieving the optimum value for label space expansion ratio requires tuning the spreading code weight to a specific value.

In the next chapter, we will establish a mathematical model for the packet throughput of OC-GMPLS core switching. The use of packet throughput as a performance measure enables us to optimize the system performance taking into consideration the tradeoff between increasing the number of multiplexed flows and the increase in the error rate on the OCDMA communication channel caused by multiplexing a larger number of flows.

Chapter 4

Throughput Analysis for

OC-GMPLS Core Switching

Chapter 3 has provided a detailed architecture, called OC-GMPLS, for utilizing OCDMA as a switching mechanism in GMPLS networks at the core. The main objective of this chapter is to present a mathematical model for the throughput of OC-GMPLS switching, which provides a quantitative measure for the performance taking into consideration the effect of the physical layer.

The flow of this chapter is as follows. Section 4.1 is a brief introduction. Section 4.2 provides a description of the optical transmission physical layer used in our proposed OC-GMPLS architecture as well as the major assumptions used in our analysis. In Section 4.3 we start by deriving the multiple access interference (MAI) bit error rate (BER) of OCDMA. The resulting BER is used in Section 4.4 to deduce a mathematical framework for the throughput of the OC-GMPLS switching architecture proposed in Chapter 3. In Section 4.5 the numerical results are calculated using the derived mathematical model. We conclude the chapter by Section 4.6.

4.1 Introduction

The OC-GMPLS switching architecture promises a significant increase in the core network switching capacity by allowing a sub-wavelength all-optical switching capability at the network core. As shown in Chapter 3, OC-GMPLS increases the all-optical switching granularity at the network core. Switching capacity by itself is not sufficient to quantify the performance of OC-GMPLS systems. An essential and important measure of switching performance is the switching throughput.

We develop a mathematical model for the switching throughput that incorporates the physical layer effects on the network layer performance by expressing the throughput as a function of the physical layer bit error rate, as well as the network traffic parameters such as the number of users and the packet length. We use our developed analytical model to evaluate the OC-GMPLS system versus prior DWDM systems, while explaining the fundamental tradeoffs between performance and capacity in OC-GMPLS network designs.

The developed mathematical model serves as a framework for a deeper understanding of the cross-layer effects on the overall switching performance. It provides a tool that can be used to fine-tune the different physical layer parameters (such as the number of codes, or the code weight) in conjunction with the network layer parameters (such as the number of users and the packet length) in order to optimize the network performance.

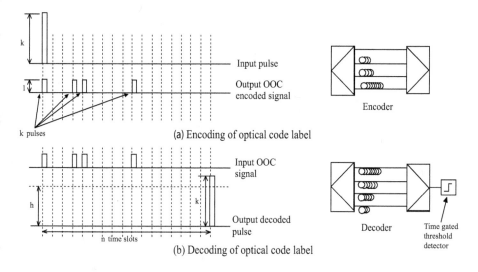

Figure 4.1: Encoding and decoding of optical code labels.

4.2 OC-GMPLS Physical Layer

We assume a physical transmission layer that employs OOK digital modulation for sending data over optical carriers. The optical spreading is performed using time domain OOC. The spreading and de-spreading of the optical signals are performed via ODLC encoders and decoders as described in Chapter 2.

While the detailed operation and design of ODLC units is given in Chapter 2, we restate here the properties of encoding and decoding units that affect our analysis. The output of the ODLC encoder is a train of On-Off Keying (OOK) digitally-modulated pulses with a unit energy and with a pattern that produces a time encoded Optical Orthogonal Code (OOC) signal with length n and code weight k as shown in Figure 4.1(a).

The decoder, shown in Figure 4.1(b), has the same structure as the encoder with the delay pattern selected such that the effect of the encoder is reversed in order to restore the original signal. The decoder also has an associated *optical time gated threshold detector*, which admits light only at a certain chip location if the total energy of the pulse is above a certain value. The use of the threshold detection device ensures that the low amplitude pulses that are uncorrelated with the decoder delay pattern are blocked from causing background noise to the correlated pulse. The device also helps against dark current effects, which are inherent in any optical detection system [70].

In order to focus on the effect of multiple access interference (MAI) due to flow multiplexing on the performance of the proposed OC-GMPLS method, the analysis presented in this section assumes an ideal channel that has zero dispersion, zero loss, and zero noise. We will also assume that all the optical components are ideal. The splitters are also assumed to divide the power equally among the different branches without any losses or distortion. Furthermore, the optical lightwaves at the couplers are assumed to be incoherent, hence, they will be recombined through simple addition of pulse energies (we neglect any electromagnetic interference effects). Another important assumption is that the encoding and decoding of the spread spectrum optical signals are performed in a chip-synchronous manner. This means that time elapses in steps of T_c, where T_c is the duration of one time chip and the transmitter and receiver clocks are perfectly synchronized at the chip level. This last assumption means that the performance analysis we give here is the worst case performance, i.e., a lower bound on the performance. This is due to the fact that MAI induced BER is at its maximum value for a chip-synchronous OCDMA system [66].

4.3 Bit Error Rate for OOK OCDMA

A key parameter that determines the transmission mechanism quality is the probability of error (P_e). To estimate the optical network overall throughput, we need to first calculate the probability of error, then derive the network throughput.

A decoding error due to MAI in an appropriately functioning[5] OOK OCDMA system happens if h or more pulses from one or more of the $N - 1$ non-intended users occur at one or more of the marked locations of the spread pulses for the intended user while the intended user is not generating spread pulses.

Assuming that a given optical source is using the code label C_i, where $i \in \{1, 2, \ldots, N\}$, and defining the spread code length to be n chips, the probability that a pulse exists at time-slot T in code C_i given that its source is active equals $\frac{k}{n}$. On the other hand, the probability that this pulse will overlap a marked pulse location from another code $C_{j \neq i}$ is equal to the probability that code C_j can have a pulse at time-slot T, which is again equal to $\frac{k}{n}$. This leads to a probability of a single chip interference (overlap) between any two codes at time-slot T equal to $\frac{k}{n} \times \frac{k}{n} = \frac{k^2}{n^2}$.

For a family of spreading codes that minimize the cross correlation (such as Optical Orthogonal Codes), a maximum of a single chip overlap is allowed between two different

---

[5]Appropriate functionality means that $h \leq k$, which guarantees correct detection for the intended user if no interference happens.

codes C_i and C_j where $i \neq j$. Consequently, the probability (p_r) of an interference to occur during a period of n time-slots can be expressed as follows:

$$p_r = \sum_{i=1}^{n} \frac{k^2}{n^2} = n \times \frac{k^2}{n^2} = \frac{k^2}{n} \ . \tag{4.1}$$

In Appendix A we show that $k^2 \leq n$ for a *non-trivial OOC*, which is a necessary condition for the expression $\frac{k^2}{n}$ to be a valid probability expression.

A given source using code C_i can be in one of two states (since there is no way of distinguishing the idle state from the zero transmission state) with the following probabilities:

- Transmitting a logical '1' with probability p_1.

- Transmitting a logical '0' with probability $p_0 = 1 - p_1$.

In order for a pulse at time-slot T to exist, its source must be transmitting a logical value of '1'. Therefore, the probability of interference p_i is given by:

$$p_i = p_1 \times p_r = p_1 \frac{k^2}{n} \ .$$

The probability of no interference is given by the complementary probability

$$1 - p_i = 1 - p_1 \frac{k^2}{n} \ . \tag{4.2}$$

From the previous argument, one can deduce that the event I_i that a user i out of the

$N - 1$ remaining users in the system interferes with the intended user has a probability $P(I_i)$ that can be expressed as

$$P(I_i) = (1 - p_1\frac{k^2}{n})\delta(I_i) + (p_1\frac{k^2}{n})\delta(I_i - 1) \, ,$$

where $i \in \{1, 2, \ldots, N - 1\}$. Accordingly, the event I of interference resulting from all the $N - 1$ remaining users in the system is given by:

$$I = \sum_{i=1}^{N-1} I_i \, ,$$

which has a probability $P(I)$ that is given by the convolution of the probabilities of the $N - 1$ independent and identically distributed (i.i.d.) random variables I_i. Hence,

$$P(I = \sum_{i=1}^{N-1} I_i) = \sum_{i=0}^{N-1} \binom{N-1}{i} \left(p_1\frac{k^2}{n}\right)^i$$
$$\times \left(1 - p_1\frac{k^2}{n}\right)^{N-i-1} \delta(I - i) \, , \tag{4.3}$$

where $\delta(.)$ is the dirac-delta function.

Assuming a threshold value of $h \leq k$ (the code weight), a detection error happens if the following combined event occurs:

$$\boxed{\text{source sending '0'} \bigcap \text{number of interferers} \geq h}$$

Note that our system cannot have errors caused by mistakenly detecting a logical '0' from a source sending a logical '1', because a logical '0' has no signal.

Therefore, the probability of error P_e is given by:

$$P_e = (1 - p_1)P(I \geq h)$$

$$= (1 - p_1) \left(\int_h^\infty P(I)dI \right) . \tag{4.4}$$

Substituting (4.3) into (4.4), we get

$$P_e = (1 - p_1) \sum_{i=h}^{N-1} \Phi(i, N) , \tag{4.5}$$

where $\Phi(i, j)$ is defined by:

$$\Phi(i, j) = \binom{j - 1}{i} \left(p_1 \frac{k^2}{n} \right)^i \left(1 - p_1 \frac{k^2}{n} \right)^{j-i-1} . \tag{4.6}$$

Equation (4.5) can be compared to the optical disk analysis provided in [66] which provides the same performance equations.

On the other hand, if $h > k$ the system will not be capable of detecting the correlated (intended) code, because the threshold will be higher than the combined energy of the de-spread pulse. This case can be divided into two subcases based on the value of N. When $(N - 1) + k \geq h$, an error in this case will occur if

> (source sending '0' $\bigcap$ no. of interferers $\geq h$) $\bigcup$
>
> (source sending '1' $\bigcap$ no. of interferers $< h - k$)

The theoretical probability of error in this case can be shown to be

$$P_{e|(N-1)+k\geq h>k} = (1 - p_1) \sum_{i=h}^{N-1} \Phi(i, N) + p_1 \sum_{i=0}^{h-k-1} \Phi(i, N) \, . \tag{4.7}$$

The second case is when $(N - 1) + k < h$. In this case the system will always detect a '0' signal and the error in this case is simply the probability of the source sending '1'.

$$P_{e|(N-1)+k<h>k} = p_1 \, . \tag{4.8}$$

Practically speaking, however, the system in both cases is not functioning because it can not detect the intended signal.

Another observation is that when $k \geq N$, h can be selected such that (3.1) is satisfied. In this case the receiver will reject all Multiple Access Interference (MAI) effects and we have $P_{e|N\leq h\leq k} = 0$. This, however, comes at the cost of increasing the code length n in order to maintain the same number of orthogonal codes N as can be deduced from (3.3). The increase in n means higher bandwidth per flow and the increase in the code weight k means that each flow will have to transmit higher energy per bit.

Collectively, the combined expression for P_e is given by:

$$P_e(N) = \begin{cases} (1 - p_1) \sum_{i=h}^{N-1} \Phi(i, N) & h \leq k \quad , \quad h < N \\ \\ P_{e|(N-1)+k \geq h > k} & h > k \quad , \quad h \leq (N-1) + k \\ \\ p_1 & h > k \quad , \quad h > (N-1) + k \\ \\ 0 & h \leq k \quad , \quad h \geq N \end{cases}$$
(4.9)

The optimum performance P_e^* (highest rejection of MAI) of the system in the case $k < N$ is achieved by setting $h = k$, which guarantees that the system is capable of detecting the intended signal if no interference happens and at the same time minimizes the MAI interference effect by rejecting up to $k - 1$ interfering users. In this case, we can rewrite (4.5) as follows:

$$P_e^* = (1 - p_1) \sum_{i=k}^{N-1} \Phi(i, N) .$$
(4.10)

Note, however, that even for an ideal system, transmission will not be error free in this case. Thus, it becomes a design tradeoff between how much multiplexing is applied versus the error threshold that must be attained. Equations (3.3) and (4.10) show that increasing k will enhance the error rate performance at the expense of limiting the maximum number of flows.

4.4 Core Switch Throughput

Although the bit error rate is a good measure of the transmission quality, it does not show how the overall network performs. In order to assess the network performance, we have to evaluate the switch throughput estimated in packets per second.

Following the model used in [81], the throughput $\mathbb{S}(N_f)$ is defined as the expected number of successful packet transmissions per unit time conditioned on the total number of input flows (N_f) offered to the system. Thus,

$$\mathbb{S}(N_f) = E\{S|N_f\} = \sum_{s=0}^{N_f} s \cdot p(s) \ , \tag{4.11}$$

where S is a random variable representing the number of successful packet transmissions, $p(s) = Prob\{S = s\}$ is the probability density function of the random variable S, and N_f is the number of active input flows offered to the system. Let the probability of a successful (correct) packet transmission be P_C. If we assume that packets have a fixed length of X bits and that bit errors are independent from one bit to the other, then the probability of successful packet transmission is equal to the probability of no error in a sequence of X bits, which is given by:

$$P_C = (1 - P_e)^X \ . \tag{4.12}$$

The probability density function of the random variable S (denoted by $p(s)$) is the probability that exactly s flows were successfully transmitted out of the N_f flows offered to the system. Thus,

$$p(s) = \binom{N_f}{s}(P_C)^s(1 - P_C)^{N_f-s} \ . \tag{4.13}$$

Substituting (4.12) and (4.13) into (4.11) gives

$$\mathbb{S}(N_f) = N_f(1 - P_e)^X \ . \tag{4.14}$$

4.4.1 Pure DWDM System Throughput

For a switch using wavelength DWDM switching with M wavelengths and fiber switching with L fibers, the bit error probability for channel number i can be expressed as [33]

$$P_e^{(WDM)} = \begin{cases} 0 & i \leq L \times M \\ \\ 1 & i > L \times M \end{cases} \ . \tag{4.15}$$

Substituting (4.15) into (4.14) gives us the throughput expression for a WDM system

$$\mathbb{S}^{(WDM)}(N_f) = \begin{cases} N_f & N_f \leq L \times M \\ \\ L \times M & N_f > L \times M \end{cases} \ . \tag{4.16}$$

4.4.2   OCDMA System Throughput

The throughput expression for an OCDMA system over a single fiber single wavelength channel given the number of flows F per wavelength per fiber can be achieved by substituting (4.9) into (4.14) to get:

$$
\mathbb{S}(F) = \begin{cases} F\left[1 - P_e(F)\right]^X & F \leq N \\ \\ N\left[1 - P_e(N)\right]^X & F > N \end{cases} , \tag{4.17}
$$

where N, the number of orthogonal codes, is bounded according to (3.3) and we choose its maximum value, which holds at the equality case. It must be noted that the throughput expression for a group of L fibers each carrying M wavelength channels each of which is capable of N OCDMA sub-wavelength labeling depends on the algorithm used to mutually assign the fiber, the wavelength and the code to the label according to the system total offered load N_f. For simplicity and comparison purposes, we assume a system where the total number of flows N_f is equally distributed among the different fibers and wavelengths. This means that the number of flows multiplexed over a single fiber single wavelength channel using OCDMA is given by

$$
F = \frac{N_f}{L \times M} .
$$

Consequently, the approximate throughput in this case can be given by the throughput of a single fiber single wavelength channel multiplied by the number of channels. This is given by

$$\mathbb{S}(N_f) = L \times M \times \mathbb{S}(F) \tag{4.18}$$

4.5 Numerical Results

The same numerical setup and assumptions as those used in deducing the numerical results in Chapter 3 are used in producing the numerical results here for the bit error rate and the switch throughput. In addition to these assumptions, we also assume that the data source is symmetric producing a '1' or a '0' bit with equal probabilities resulting in $p_1 = \frac{1}{2}$. It must be noted that in the semi-log figures for P_e, we use a very small value of around 10^{-15} instead of $-\infty$ to represent $\log(0)$. This helps overcoming the scaling problems when plotting our results and at the same time illustrates that a MAI-free system still have some errors due to channel and components imperfections.

In Figure 4.2, the influence of changing the number of multiplexed flows and the code weight on the transmission BER is demonstrated. It can be seen from the figure that there are several regions over which the BER behaves differently with changes in F and k. Figure 4.3 shows the different regions of BER behavior and how they are affected by the relative relations between different system parameters. The first region is the one where the number of multiplexed flows is less than the maximum number of orthogonal codes defined by (3.3). In this region, increasing F results in higher probability of error. It must be noted, however, that in order to maintain the orthogonality condition, F is not allowed to be larger than N. This is also illustrated by the loading line defined in

Figure 3.5 in Chapter 3. When F reaches the value of N, it is set fixed to this value. This

is the reason of the saturation seen in Figure 4.2. Changing the code weight, on the other

hand, has a different and more dramatic effect on the system performance. This is due

to two main reasons. First, the value of k relative to the value of the optical threshold

h defines the operational region of the system as described in Chapter 3. Second, the

value of k affects N, which eventually limits the number of flows in the system. From

Figure 4.3 it can be seen that for the region $k < h$, the system has a BER of about 0.5

due to the fact that it cannot detect the intended signal. Increasing the code weight

moves the system into the appropriate operational region where an increase in k would

increase the system BER due to the increase in the number of pulses per bit. Further

increase in k results in N becoming smaller till it starts reducing the number of flows in

the system. At this point an increase in k in fact decreases the BER due to the required

reduction in the number of multiplexed flows. This continues until the point where N

and hence F becomes lower than the optical threshold, at which point the BER drops to

zero and the system is error-free.

In Figure 4.4, the optimum value of BER at $h = k$ under maximum switching capacity

condition given by the equality case in (3.3) is plotted against code weight and code

length. It can be seen that increasing k dramatically reduces the BER. This effect is

due to the increase in h because of the optimality condition as well as the decrease in N

due to the maximum capacity condition. Increasing h combined with reducing N rapidly reduces the MAI error effect till the point where it is completely eliminated. On the other hand, increasing n slightly increases the BER due to the increase in N caused by the maximum capacity condition again. The values of N that correspond to the value of P_e (shown in Figure 4.4) are shown in Figure 3.8, while the corresponding throughput values are shown in Figure 4.9.

Figure 4.5 demonstrates how a single wavelength channel can be utilized to carry a number of sub-wavelength multiplexed or labelled flows that is equal to the number of DWDM channels carried over a single fiber standardized by the ITU-T DWDM grids while achieving a throughput (in terms of the average number of successful transmissions) very close to that achieved by the DWDM system. The reason that the throughput is slightly less than the number of multiplexed flows is because of the MAI induced errors. Multiplying the number of sub-wavelength labels by the number of wavelength channels that can be multiplexed on a fiber and multiplying the result by the number of fibers gives an average estimate for the size of the expanded all-optical core label mapping space achieved using our method with marginal loss of performance.

Figure 4.6 in conjunction with Figure 4.7 and Figure 4.8 show how the core switch throughput is affected by increasing the code weight k as well as the number of offered flows per wavelength per fiber F according to (4.17). Figure 4.7 can be thought of as cross-sections of Figure 4.6 over planes parallel to the $F - \mathbb{S}$ plane at different values of k. Similarly, Figure 4.8 is composed of cross-sections to Figure 4.6 over planes parallel to the $k - \mathbb{S}$ plane at different values of F. It can be seen from these figures that the throughput has the same optimum value whether it is calculated against k or F. However, an important difference between the two calculation approaches is that the value of F that achieves the maximum throughput depends on the selected value of k, while the value of k that maximizes $\mathbb{S}$ is independent of the selected number of flows F. This is due to the maximum capacity condition used to define the loading curve. The behavior of the throughput shown in Figure 4.7 where it increases up to a maximum value before it starts to decline and settles at a lower value can be attributed to the higher error rate value at the points after the maximum throughput point. Essentially, the higher error rate overcomes the rise in throughput due to the increase in number of offered flows, and causes the overall achievable throughput to decrease until the point where the system reaches the maximum number of allowable flows and further increase in offered load is not possible.

Figure 4.9 reconfirms that the throughput conditioned on optimum threshold and maximum capacity is more sensitive to changes in the code weight than it is to changes in the code length. This is evident by the fact that changes in k cause a non-linear change in $\mathbb{S}$, while changes in n affect $\mathbb{S}$ in a linear manner.

It is evident from Figure 4.10 that the throughput is exponentially decaying with increasing packet length for small values of k. This behavior, however, vanishes when k increases because the BER approaches zero, resulting in a throughput that is independent of the packet length.

The trajectory of the maximum throughput achievable $\mathbb{S}_{max}$ measured against the number of flows as well as the trajectory of the corresponding values of code weight are shown in Figure 4.11. Similarly, the trajectory of the maximum throughput achievable S_{max} measured against the code weight as well as the trajectory of the corresponding number of multiplexed flows are shown in Figure 4.12. Observe that there is an optimum value of about 30 for the throughput and the corresponding number of flows is given by $F = 37$ while the corresponding code weight is given by $k = 5$. Both values can be achieved through either of the trajectory with k or the trajectory with F. The reason that the maximum achievable throughput is less than the number of flows is the fact that there is a non-zero value for the probability of error at the maximum throughput point.

Figure 4.13 shows how these two trajectories move on the surface of S_{max} as a function of F and k. It can be seen from the figure that the two trajectories start and finish at different values for S_{max} however they agree at several points around the optimum value and at the optimum value as well. The two trajectories can be thought of as the path that a search algorithm uses to find the maximum value in a matrix of throughput values with rows representing k and columns representing F. If the search algorithm spans rows, it follows the trajectory against F (solid line), otherwise it follows the trajectory against k (dotted line) and both will lead to the same maximum throughput value.

The previously mentioned scenarios show that there is an optimum choice for the number of orthogonal codes (maximum number of flows) N that can be multiplexed on a single wavelength channel using OCDMA. With the parameters used in this numerical analysis, up to $F = 37$ flows can be multiplexed over the same wavelength.

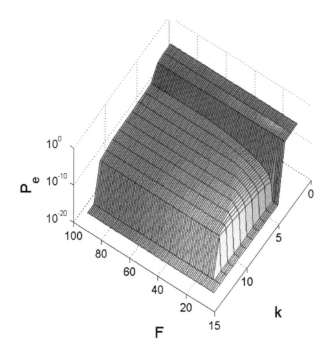

Figure 4.2: Effect of number of flows and code weight on BER of OOK OCDMA transmission ($h = 5$).

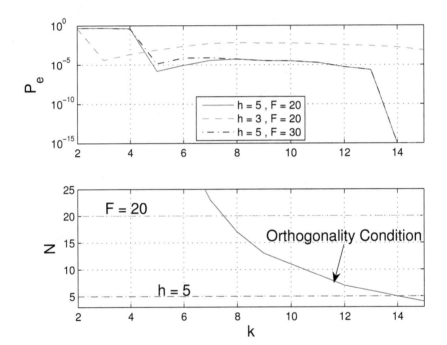

Figure 4.3: BER operational regions for OOK OCDMA transmission ($X = 1024$ bits).

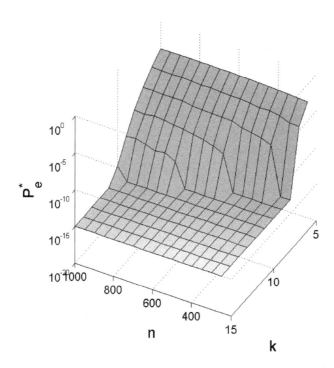

Figure 4.4: Optimum BER as a function of code length and code weight.

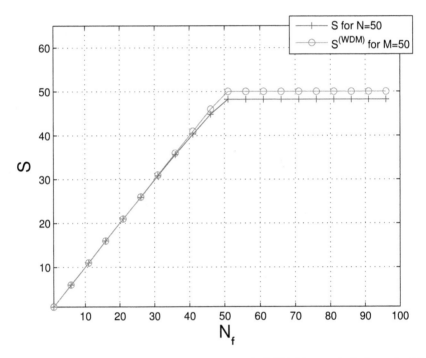

Figure 4.5: Throughput of a single sub-wavelength OCDMA switched channel with 50 orthogonal codes as compared to the throughput of 50 DWDM wavelength switched channels (X = 1024).

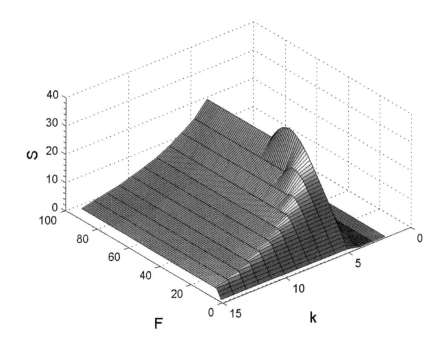

Figure 4.6: Throughput of OC-GPMLS core switching as a function of the number of sub-wavelength multiplexed flows and code weight.

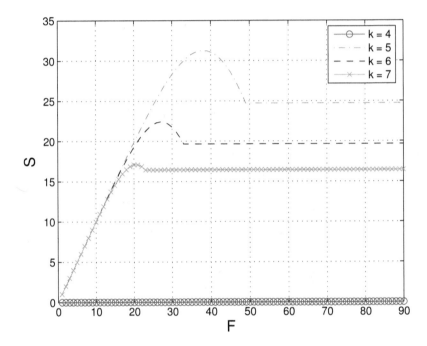

Figure 4.7: Effect of the number of sub-wavelength multiplexed flows on the throughput of OC-GMPLS core switching.

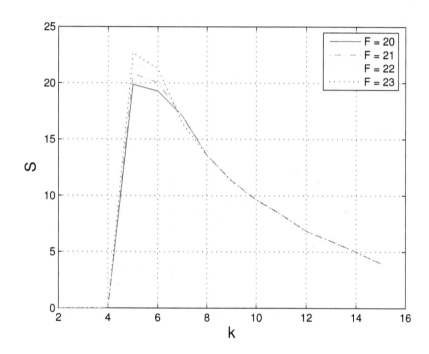

Figure 4.8: Effect of code weight on the throughput of OC-GMPLS core switching.

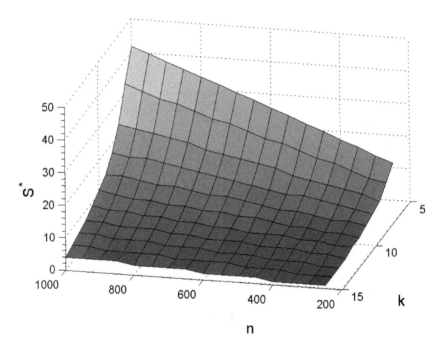

Figure 4.9: Effect of code weight and code length on the throughput of OC-GMPLS core switching at optimum optical threshold value ($h = k$) under the maximum switching capacity condition.

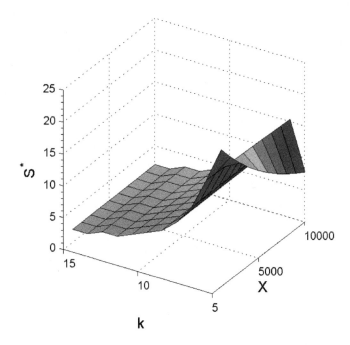

Figure 4.10: Effect of code weight and packet length on the throughput of OC-GMPLS core switching at optimum optical threshold value ($h = k$) under the maximum switching capacity condition.

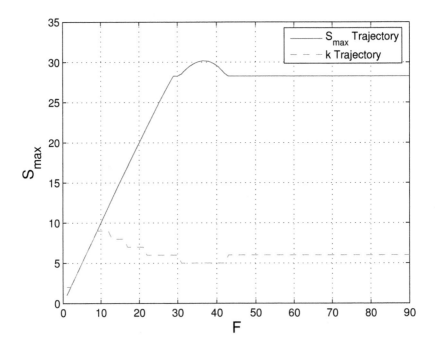

Figure 4.11: Trajectory of maximum throughput of OC-GMPLS core switching against number of multiplexed flows for code length = 900 chips and packet length = 3000 bits using optimum optical threshold value $h = k$.

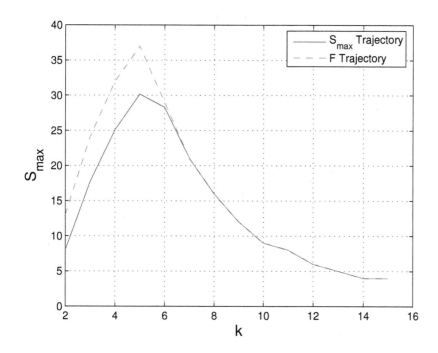

Figure 4.12: Trajectory of maximum throughput of OC-GMPLS core switching against code weight for code length = 900 chips and packet length = 3000 bits using optimum optical threshold value $h = k$.

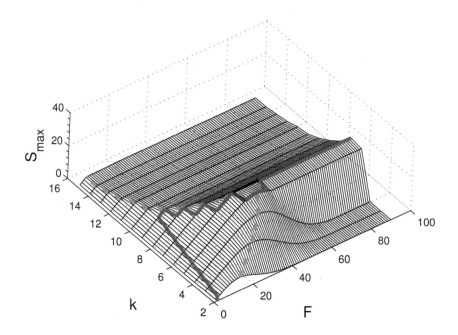

Figure 4.13: 3-D trajectories of maximum throughput through number of multiplexed flows (solid thick line) and code weight (dotted thick line) for code length = 900 chips and packet length = 3000 bits using optimum optical threshold value $h = k$.

4.6 Concluding Remarks

Using OOK in association with Fiber Tapped Delay Line correlators to implement OC-GMPLS and assuming a uniform label assignment algorithm, we were able to provide a mathematical analysis for the core switch performance of OC-GMPLS. Our mathematical analysis started by deducing an expression for the MAI induced BER, which we have used to find an expression for the average switch throughput as a function of physical layer parameters such as the BER, code length and code weight as well as network layer parameters such as the number of multiplexed users and the average packet length.

Analyzing the switch throughput performance, we were able to show that there is a design tradeoff between maximizing the throughput of the core switch and increasing the maximum number of multiplexed flows on the same fiber and wavelength channel. Our numerical results show that there is an optimum point which maximizes both objectives. We have found the optimum point using a simple search algorithm that finds the maximum value in a matrix of the throughput values via sequential search either through columns first or rows first.

The mathematical analysis reveals that the system has different operational regions. Each of the operational regions has different characteristics. Some of these operational regions are not practically useful and should be avoided such as the case where the optical threshold is larger than the code weight ($h > k$). Other different regions fit particular application requirements. The MAI error free region is useful where high transmission reliability is needed. It guarantees zero MAI errors while it does not necessarily achieve the maximum system capacity in terms of throughput and number of multiplexed flows. On the contrary, when throughput and capacity are all that matters, reliability can be sacrificed by multiplexing more users while allowing for higher BER. Our analysis shows, however, that this trade-off has a limit after which adding more users (flows) to the system would degrade both the BER (reliability) and the average throughput (capacity).

Throughout the work in this chapter, we have assumed the use of On-Off Keying (OOK) as a digital modulation technique. Although OOK is the most common modulation technique in optical communications, it introduces several practical challenges. In the next chapter, we address the practical challenges introduced by using OOK and propose a modulation method that can overcome these issues while delivering a BER performance that asymptotically approaches the BER performance of OOK.

Chapter 5

Chip-Level Modulated BPPM

Fiber-Optic CDMA

In Chapter 4, we have discussed the throughput of core switching using OC-GMPLS with OOC time encoded OCDMA and OOK digital modulation. In this chapter, we will focus on the digital modulation aspect of OCDMA transmission systems. Our aim is to propose a modulation technique that achieves a BER close to that of OOK while overcoming the synchronization and source activity detection problems introduced by OOK. Our proposed modulation method is called Chip-Level Modulated Binary Pulse Position Modulation (CLM-BPPM).

The rest of this chapter is organized as follows. We provide a brief introduction in Section 5.1. In Section 5.2 we describe the operation of the proposed CLM-BPPM modulation technique, its different variations and how it can be implemented using an all-optical system. Section 5.3 provides a channel model for CLM-BPPM transmission as part of a framework for modeling the system performance mathematically. In Section 5.4, we derive a generalized expression for the BER of the CLM-BPPM due to Multiple Access Interference (MAI) under ideal channel conditions. We use the generalized BER mathematical framework to deduce an expression for the BER of a variation of the CLM-BPPM that uses absolutely timed detection and '1'-value first chip conflict resolution. Finally, Sections 5.5 and 5.6 provide numerical results for the performance of the CLM-BPPM followed by concluding remarks on the work presented in this chapter.

5.1 Introduction

Classical optical transmission systems use the OOK binary modulation scheme to send data over the power of optical carriers. With OOK it is difficult for a receiver to perform clock recovery or to differentiate between active sending '0' and idle states of the source. An alternative scheme called Binary Pulse Position Modulation (BPPM) has been proposed as a binary modulation scheme for optical fiber transmission [45,82]. Pulse Position Modulation (PPM) techniques guarantee a transition at least every bit duration, which simplifies the clock recovery at the receiver side. PPM also guarantees that an optical signal is transmitted for any value including a '0' bit value as long as the source is active. This means that the only case where there will be no light detected at the receiver for an entire bit duration is when the source is idle. This simplifies source activity state detection by the receiver. PPM modulation techniques proposed in the literature are all based on performing the position modulation at the bit level then spreading the modulated signal (M-symbol pulse) using the optical spreading code. This results in disturbing the spreading code correlation properties over the bit duration. This disturbance can be demonstrated by observing the scenario where a '1'-bit spread using OOC and bit level binary PPM is received with a half bit time-shift. This time-shifted bit can be interpreted by another receiver as a '0'-bit interference, which means that active users sending both '1'-bits and '0'-bits contribute to '0'-bit detection errors. The same scenario applies to the '1'-bit detection case. This degrades the system performance unless strict bit level synchronization between different users is applied, which is practically difficult.

In this chapter, we propose the use of a modulation mechanism that uses the value of the optical bit to perform BPPM of the chip pulses of the optical spread signal generated using an Optical Orthogonal Code (OOC) for time spreading. We call this mechanism Chip-Level Modulated Binary Pulse Position Modulation (CLM-BPPM). This mechanism preserves all the properties of the classical bit level PPM with the added advantage of maintaining the spreading code correlation properties over the entire bit duration. We propose several variations of the mechanism based on detection and chip conflict resolution methods.

In our effort to analyze the performance of the proposed scheme, we derive a general mathematical framework that can be applied to any optical time spread OCDMA system. Our mathematical framework takes into consideration for the first time the effect of the source activity average statistics on the bit error rate of the system. We apply our mathematical framework to one variation of the CLM-BPPM mechanism and show that the performance of our proposed system is very close to the OOK system with equal asymptotic BER of the two systems at full source activity conditions.

5.2 Proposed Chip-Level Modulated BPPM

(CLM-BPPM)

Now, we present a novel OCDMA incoherent transmission method using the principles of BPPM for the binary modulation case, which can be readily extended to M-ary Pulse Position Modulation (M-PPM) for the M-ary modulation case. The BPPM can be viewed as a generalized version of the Manchester coding scheme.

Our proposed OCDMA transmission method is called Chip-Level Modulated Binary Pulse Position Modulation (CLM-BPPM) OCDMA. CLM-BPPM OCDMA has several variations (options) depending on the detection methods used by the receiver and the chip conflict resolution methods (i.e., the methods used to resolve the case where a '0' chip and a '1' chip interfere). We have three detection methods at the receiver: *Absolutely Timed Detection*, *Untimed Detection*, and *Differentially Timed Detection*. The chip conflict resolution can be one of three methods: *'X'-Value First*, *Majority with Random Resolution*, or *Majority with 'X'-Value First*. Later in this section, we will describe the functionality of each of these techniques.

5.2.1 Modulation Method

The modulation method refers to how the '1' and '0' bit values are represented at the optical CDMA signal level. CLM-BPPM modulation is described as follows (see Figure 5.1). Each chip duration T_c is divided into two slots. If a source is generating a binary '1', it will produce a short pulse of duration $T_c/2$ that occupies the second half of T_c. This pulse will go through the different fiber delay lines of the ODLC encoder to generate the spread signal. The spread signal will consist of k pulses each with duration $T_c/2$ at the second halves of the marked chip locations determined by the spreading code within the n chip location occupying a full bit duration T_b. At the receiver side, after the de-spreading operation is performed in the ODLC decoder, the detection device will sense a low-to-high transition in the middle of the chip duration, which represents a binary '1' in our BPPM scheme. On the other hand, if the source is generating a binary '0', it will produce a pulse that occupies the first half of the chip duration, which will be seen as a high-to-low transition by the detector. This high-to-low transition represents a binary '0' in the BPPM scheme. If there is no low-to-high nor high-to-low transition detected by the receiver within a chip duration, the source is assumed to be idle 'IL' during this chip[6]. For this transmission method we use an OOC spreading code family composed of N orthogonal codewords each has a code length n and code weight k. Accordingly, the relation between T_c and T_b is given by:

$$T_c = \frac{T_b}{n} \ .$$

This modulation method is the common base to all variations that we propose in this section. It provides for an easier clock recovery mechanism, because at least k transitions

---

[6]Note that this is different from a source being idle for a bit duration (i.e., sending no data) in which case it will have 'IL' for all chips over an entire bit duration.

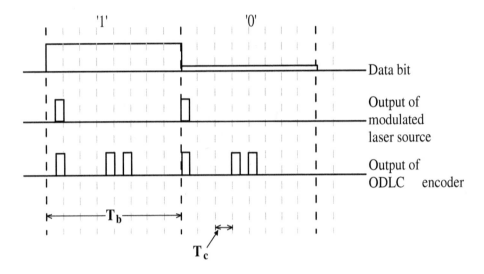

Figure 5.1: CLM-BPPM modulation method.

every bit duration are guaranteed when the source is active whether it is sending a '0'

or a '1'. It is also possible for the receiver to detect idle source states when it detects

no transitions within an entire bit duration. The receiver for this method uses a passive

correlator detection architecture utilizing matched filter optical correlators based on fiber

tapped delay lines. The receiver, however, performs photon counting over individual chip

durations in order to detect the bit value (in this case it counts the photons in the signal

resulting from the optical correlator). In this sense our receiver can be considered as a

type of chip-level based detector.

5.2.2 Transceiver Architecture

Although we propose several alternative methods for the receiver operation, our proposed transceiver architecture is not affected by these variations. The architecture of the CLM-BPPM OCDMA transceiver is shown in Figure 5.2. As can be seen from the architecture, in order to be able to modulate the chip with the BPPM data, we use a modulated laser source, which is controlled by the digital data source. The laser produces a pulse that has a duration $T_c/2$ and its timing within the chip duration is controlled by the data source to represent the BPPM scheme described earlier. The laser source is connected to the ODLC encoder, which spreads the signal in the time domain by producing lower energy replicas of the pulse at the positions marked for pulse transmissions corresponding to the locations of 1's in the OOC used. The output goes into port 1 and out of port 2 of the 3-port circulator C3. It then couples through a 50:50 optical splitter to the fiber line via port 1 and port 2 of both circulators; C1 and C2.

The receiver part is connected through port 3 of C3, which has port 2 connected to the optical splitter. The signal on the fiber follows the paths shown with grey and dotted arrows through circulators C1 or C2. The signal is reverse coupled into port 2 of C3 through one of the splitter's output ports and an Erbium Doped Fiber Amplifier (EDFA). The signal leaves the circulator through port 3 and goes into the ODLC decoder. The ODLC decoder perform despreading, optical threshold detection and optical timed gating (sampling) of the correlated received optical signal. It feeds the resulting optical signal into the optical detector. The detector changes the signal from the optical format

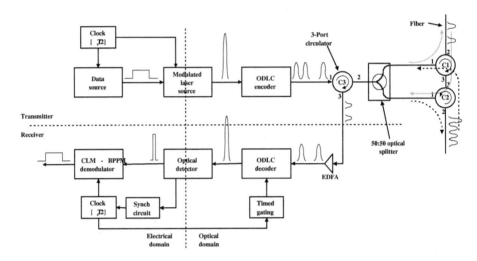

Figure 5.2: CLM-BPPM OCDMA transceiver architecture.

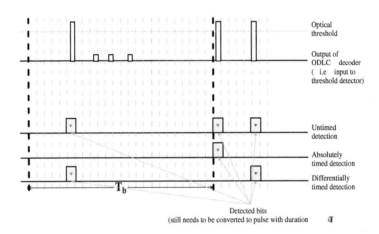

Figure 5.3: Illustration of different detection methods.

into the electrical format and feeds it into an electric CLM-BPPM demodulation circuit. The CLM-BPPM decoder decodes the signal into its original bit format. The optical detector is also connected to a synchronization block, which extracts clock phase and frequency from the received signal. The local clock is used to derive both the CLM-BPPM demodulator and an optical timed gate controller, which controls the timing of the optical timed gating operation in the ODLC decoder.

5.2.3 Detection Methods

After a signal is modulated, spread, and sent by the transmitter, the receiver is required to despread the signal then perform a detection operation in order to convert the signal from its binary modulated form into its bit logic values of '1' or '0' bits. The latter operation is referred to in our work as the *detection method*. In the following we introduce three different detection methods (Figure 5.3) based on the correlator receiver architecture with optical threshold detection.

Absolutely Timed Detection is based on allowing detections of chips only at certain chip locations. For the correlator receiver, the detection is performed only at the chip next to the last chip of each bit duration T_b. Using this method, any erroneous chip value within T_b that exceeds the threshold due to MAI effects will be masked out by the detection process except if this situation happens at the intended detected chip. This will greatly reduce the error rate of the receiver. However, this type of detection is greatly dependent on continuous synchronization between the transmitter and the receiver. Any drift or loss of synchronization will have a large impact on the system performance.

Untimed Detection is a detection method where the optical receiver completely relies on the value of the chip regardless of its position within the bit duration. This means that as soon as the photon count over a chip duration exceeds the threshold value, the detector detects a bit value (either '1' or '0' depending on the BPPM demodulation). This happens even if two chip pulses with energies higher than the detection threshold occur within the same bit duration T_b. Accordingly, if these two pulses occur within the same bit duration the receiver will detect two bits within one bit duration. This will cause an erroneous detection, however, implementing this method is easier since no bit level timing information needs to be communicated between the transmitter and the receiver.

Differentially Timed Detection requires that two chips exceeding the threshold value can be detected as a '1' or '0' bit if and only if they are separated by a time duration greater than or equal to one bit duration T_b. This means that if two chips with energy values greater than the threshold are received with time separation less than T_b, the second chip will be ignored. This method is less strict in terms of synchronization requirements than the absolutely timed method. It is, however, expected to perform better than the untimed method due to the fact that it omits an erroneous situation that is detected in the untimed method as a legitimate bit.

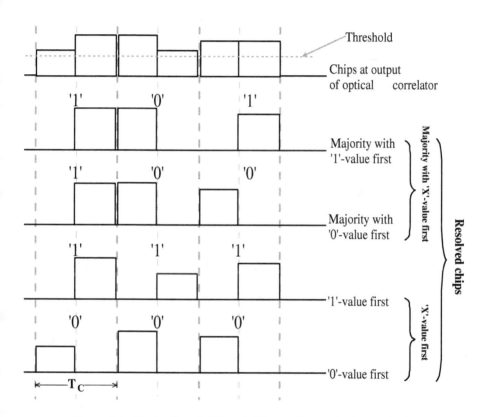

Figure 5.4: Illustration of different chip conflict resolution methods.

5.2.4   Chip Conflict Resolution Methods

Due to the fact that in our BPPM modulation scheme both '1' and '0' bit values result in an optical signal with equal energy values, situations may happen where the MAI can cause a '1' and a '0' chip pulse to interfere at the receiver side. The conflict resolution methods aim at resolving these conflicts once they happen based on a strategy that will allow a '1' or a '0' chip value to be detected rather than assuming a detection error. This way the bit error rate of the system can be reduced. We propose here three different methods to resolve chip conflicts (see Figure 5.4).

'X'-Value First method is based on preassigning a priority to '1' chip value or '0' chip value (the 'X'-value, where 'X' $\in$ {'1','0'}). Meaning that if a chip conflict happens, it is always detected as either a '1' or a '0' based on the preassigned priority. This results in a simpler receiver architecture, since no comparison process is required. However, this method is biased towards one of the chip values (the value that has the priority) and will have different performance results depending on the pattern of the bits generated by the transmitter.

Majority with Random Resolution performs counting of the photons over the two halves of the chip duration. If the majority is in the second half of the chip duration, the chip is detected as a '1', otherwise it is detected as a '0'. In case of equality, a random selection with probability p_s is performed to decide on a '1'. Otherwise, a '0' decision is taken. This method provides a fair treatment for '1' and '0' valued chips, but it requires photon counting over each half of the chip duration separately then a comparison.

Majority with 'X'-Value First is the same as *Majority with Random Resolution* with the difference that in case of equality, preference is given to 'X' chip value (either '1' or '0' depending on the value chosen for 'X' in the scheme).

5.3 MAI Channel Model

In order to focus on the effect of user multiplexing on the performance of the proposed CLM-BPPM method, we assume an ideal channel that has zero dispersion, zero loss, and zero noise. Generally speaking, a nominal value of acceptable BER in a MAI interference-free optical transmission system is in the order of 10^{-12}. In our analysis we consider MAI interference levels that produce higher BER values. This allows us to focus on the MAI effect and neglect the optical noise effects. For noise models in optical communications and their effects on BER performance the reader is referred to [83,84]. We will also assume that all the optical components are ideal. The splitters are assumed to divide the power equally among the different branches without any losses or distortion. Furthermore, perfect synchronization between the transmitter and the receiver at the chip duration level is assumed. The optical lightwaves at the couplers are assumed to be incoherent, hence, they will be recombined through simple addition of pulse energies (we neglect any electromagnetic coupling or interference effects between the recombined lightwaves).

In our system the optical detector detects the data bit based on photon counting over every chip duration at the output of the ODLC. In an ideal system that has zero MAI effects, the output of the ODLC decoder after the threshold detector should match

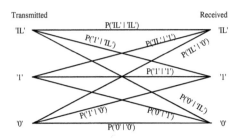

Figure 5.5: General CLM-BPPM OCDMA channel model.

chip-by-chip to the input from the laser source to the encoding ODLC with a time shift

equal to T_b (see Figure 2.4 in Chapter 2).

When the optical threshold detector device passes a pulse, the chip detector uses the

pulse position within the chip duration T_c to decode the bit as explained earlier. If the

receiver does not detect a chip pulse during a chip duration it assumes an idle chip. A

chip can have one of three states: idle ('IL'), carrying a logical one ('1') or carrying a

logical zero ('0'). Thus, the channel can be modeled as a tristate channel. Figure 5.5

demonstrates the general channel model. For each of the transmission technique vari-

ations described in Section 5.2, the conditional probabilities of receiving state j given

that state i was transmitted $\{P(j \mid i) : i, j \in \mathbb{B} = \{'0', '1', 'IL'\}\}$ takes different values,

resulting in a different probability of error.

The correct decision space $\mathbb{C} = \{(i, j) : i, j \in \mathbb{B}\}$ is defined as the set containing

all the ordered pairs (i, j) of points connected by horizontal lines in Figure 5.5, where i

represents the transmitted value and j represents the received (detected) value . This is given by:

$$\mathbb{C} = \{(`1`, `1`), (`0`, `0`), (`IL`, `IL`)\} \ .$$

Hence, the error space $\mathbb{E} = \{(i, j) : i, j \in \mathbb{B}\}$ is defined as the space containing all the pairs of points in Figure 5.5 that are not connected by horizontal arrows and is given by:

$$\mathbb{E} = \mathbb{B}^2 - \mathbb{C} \ ,$$

where $\mathbb{B}^2$ is the ordered cartesian product of $\mathbb{B}$ by itself.

Using the general channel model, the probability of error can be expressed as:

$$P_e = \sum_{\mathbb{E}} P(j \mid i) \ . \tag{5.1}$$

Due to the assumption of zero attenuation on the channel and perfect chip-level synchronization between transmitter and receiver, the probability of detecting an idle chip given that a chip was sent to represent a '1' or a '0' bit is equal to zero. Accordingly,

$$P(`IL` \mid `1`) = P(`IL` \mid `0`) = 0 \ , \tag{5.2}$$

which means that we have

$$\mathbb{E} = \{(`0`, `1`), (`1`, `0`), (`IL`, `0`), (`IL`, `1`)\} \ . \tag{5.3}$$

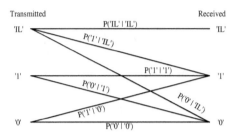

Figure 5.6: Ideal CLM-BPPM OCDMA channel model.

Figure 5.6 shows the particular channel model for an ideal CLM-BPPM scheme satisfying (5.2), where all branches with probability equal to zero are eliminated.

Depending on the modulation method, the detection method, and the bit conflict resolution method, the expression in (5.2) can have different values. The changes in (5.2) can be mainly categorized into two main categories:

- Reduction in the cardinality of the set $\mathbb{E}$ due to having $P(j \mid i) = 0$ for some pairs. This can be caused only by changing the chip conflict resolution mechanism.

- Changing the values of $P(j \mid i)$ for some elements $(i,j) \in \mathbb{E}$. This can be caused by changing any of the modulation, detection or chip conflict resolution methods.

5.4 Performance Analysis

The system performance is measured in terms of the bit error rate (or probability of error P_e) due to MAI only. In order for the analysis to be mathematically tractable, the encoding and decoding of the spread spectrum optical signals is assumed to be performed in a chip synchronous manner. This means that time elapses in steps of T_c, where T_c is the duration of one time chip and the transmitter and receiver are perfectly synchronized at the chip level. This assumption means that the performance analysis we give here is the worst case performance, i.e., a lower bound on the performance similar to that generated by [66].

In the following analysis, we consider a system of N_u users (optical sources). We assume that a given optical source numbered x (called the *intended source*) uses codeword $C_x \in \mathcal{C}$, where $\mathcal{C}$ is a family of N orthogonal codes and $x \in \{1, 2, ...N\}$. Each codeword has a code weight (number of pulses per codeword) k and code length n chips per codeword. Each of the remaining $N_u - 1$ users in the system uses a distinct codeword $\{C_y : C_y \in \mathcal{C}, y \in \{1, 2, ...N\}, y \neq x\}$

The general parameters controlling the system performance are:

1. The number of users N_u in the system.

2. The threshold value h of the optical threshold detector.

3. The number of orthogonal codes N in the OOC family.

4. The code weight k of the OOC.

5. The code length n of the OOC.

In our analysis, we will consider configurations having the following parameter constraints:

$$N_u \leq N , \tag{5.4}$$

$$h \leq k , \tag{5.5}$$

$$N \geq h . \tag{5.6}$$

The condition in (5.4) guarantees that only the effect of interference between orthogonal codes is taken into consideration (i.e., we do not consider code reuse among users). The maximum user capacity (and at the same time highest level of MAI) for the system is reached when equality holds [15, 25]. What (5.5) ensures is that the system is within its operational region (i.e., capable of detecting the intended signal). This can be explained by noticing that for $h > k$, the system will not be able to detect the intended signal at the output of the decoding ODLC since the threshold value will be larger than the total energy of the recombined chips [25]. Finally, (5.6) tells that our system is working on the region where MAI can cause detection errors, for if the number of users N_u was less than the threshold value, the total MAI energy at a single chip location will be less than the threshold value and will always be eliminated producing an error-free system [25].

5.4.1 BER Mathematical Analysis Framework

A general model for the bit error rate (BER) (or probability of error) of the system presented in this chapter starts from (5.1). Our goal is to derive generalized expressions for each of the different conditional probabilities in (5.1).

Consider a system that consists of N_u users using spread codes with code length n chips and code weight k pulses per bit. Assuming that users have independent transmissions with uniformly distributed arbitrary delays between them, there are k^2 ways, with probability $1/n$ each, that a pulse from an interfering user y using spread code C_y will interfere with a pulse location in code C_x used by intended user x. Knowing that for a family of optical spreading codes that minimize the cross correlation (such as Optical Orthogonal Codes), a maximum of a single chip overlap is allowed between any two different codes C_x and C_y where $x \neq y$, the probability (q) of an interference from user y to occur during a period of n chips (i.e., a one bit period) given that source y is active can be expressed as follows [53]:

$$q = \frac{k^2}{n} .$$

(5.7)

Assuming that a source y is active with probability P_A and transmits '1' bits when active with probability P_1, we can find the two probabilities that source y is interfering with a '1' bit value (q_1) or with a '0' bit value (q_0) as follows

$$q_1 = P_A P_1 \frac{k^2}{n}$$

(5.8)

$$q_0 = P_A (1 - P_1) \frac{k^2}{n} .$$

(5.9)

We assume that at any point in time, out of the $N_u - 1$ possible interfering sources there are l sending '1' bits and m sending '0' bits while the remaining are either idle 'IL'

or active but not interfering with the intended transmission. l and m satisfy the following

conditions: $0 \leq l \leq N_u - 1$, $0 \leq m \leq N_u - 1$ and $l + m \leq N_u - 1$. Therefore, the joint

probability of having l '1'-interferes and m '0'-interferers at any of the k pulse locations

of the intended source i is given by the multinomial distribution with parameters q_1, q_0,

$N_u - 1$, l and m

$$P(l,m) = \binom{N_u - 1}{l, m} q_1^l q_0^m \left(1 - q_1 - q_0\right)^{N_u - 1 - l - m}, \tag{5.10}$$

where

$$\binom{N_u - 1}{l, m} = \binom{N_u - 1}{l} \binom{N_u - 1 - l}{m}.$$

The number of interferers is not the only factor that governs the occurrence of errors

in an OCDMA system using OOC. The error events are greatly affected by the pattern

of the interference (i.e., the number of interferers in each of the k slots corresponding to

the positions of pulses in the intended source code C_x). We introduce two k-dimensional

vectors to represent the patterns for '1' and '0' interferences. These vectors are given by

$$\overrightarrow{\alpha}(l) = \left\{ \alpha_i(l) : \alpha_i(l) \in \{0, 1, \cdots, l\}, \ i = 1, 2, \cdots, k \right\}, \tag{5.11}$$

where $\alpha_i(l)$ is the number of '1'-interferers in a slot corresponding to the location of pulse number i in the intended source code C_x given that there are l '1'-interferers and

$$\overrightarrow{\beta}(m) = \left\{ \beta_i(m) : \beta_i(m) \in \{0, 1, \cdots, m\}, \ i = 1, 2, \cdots, k \right\}, \tag{5.12}$$

where $\beta_i(m)$ is the number of '0'-interferers in a slot corresponding to the location of pulse number i in the intended source code C_x given that there are m '0'-interferers.

By definition, the following two conditions for OOC based time spreading must apply to the interference pattern vectors

$$\sum_{i=1}^{k} \alpha_i(l) = l \tag{5.13}$$

$$\sum_{i=1}^{k} \beta_i(m) = m . \tag{5.14}$$

For different values of l and m there exists a set $\mathbb{F}_l$ of '1'-interference patterns $\overrightarrow{\alpha}(l)$ and another set $\mathbb{G}_m$ of '0'-interference patterns $\overrightarrow{\beta}(m)$ that satisfy (5.13) and (5.14) respectively. Mathematically, we define $\mathbb{F}_l$ and $\mathbb{G}_m$ as follows

$$\mathbb{F}_l = \left\{ \overrightarrow{\alpha} : \sum_{i=1}^{k} \alpha_i = l , \quad \alpha_i = 0, 1, \cdots, l \right\} \tag{5.15}$$

$$\mathbb{G}_m = \left\{ \overrightarrow{\beta} : \sum_{i=1}^{k} \beta_i = m , \quad \beta_i = 0, 1, \cdots, m \right\} . \tag{5.16}$$

Next, we deduce the joint probability $P(\overrightarrow{\alpha} \in \mathbb{F}_l, \overrightarrow{\beta} \in \mathbb{G}_m \mid l, m)$ that interference patterns $\overrightarrow{\alpha}(l)$ and $\overrightarrow{\beta}(m)$ happen given that there are l '1'-interferers and m '0'-interferers in the system. We observe that given l and m, the events $\{\overrightarrow{\alpha} \in \mathbb{F}_l \mid l\}$ and $\{\overrightarrow{\beta} \in \mathbb{G}_m \mid m\}$ are independent. Hence,

$$P(\overrightarrow{\alpha} \in \mathbb{F}_l, \overrightarrow{\beta} \in \mathbb{G}_m \mid l, m) = P(\overrightarrow{\alpha} \in \mathbb{F}_l \mid l)P(\overrightarrow{\beta} \in \mathbb{G}_m \mid m) . \tag{5.17}$$

The problem now reduces to finding the probabilities $P(\overrightarrow{\alpha} \in \mathbb{F}_l \mid l)$ and $P(\overrightarrow{\beta} \in \mathbb{G}_m \mid m)$ separately.

With l independent users, each sending a '1' bit (with k chips) and introducing exactly one chip overlap with any other user in the group (due to the use of OOC as spreading codes), the problem of finding the probability of different user interference patterns reduces to the problem of choosing l elements out of k elements with replacement, which can happen in k^l equally probable ways each with probability $1/k^l$. However, what matters in our system is the chip interference pattern itself regardless of the particular pattern of users producing it. Having said that, we notice that the number of different user interference patterns that produces the same chip interference pattern $\overrightarrow{\alpha}(l)$ is given by the multinomial coefficient

$$\binom{l}{\alpha_1, \alpha_2, \cdots, \alpha_k} = \binom{l}{\alpha_1}\binom{l-\alpha_1}{\alpha_2}\cdots\binom{\alpha_k}{\alpha_k} = \frac{l!}{\prod\limits_{i=1}^{k} \alpha_i!} .$$

Consequently,

$$P(\overrightarrow{\alpha} \in \mathbb{F}_l \mid l) = \frac{l!}{\left(\prod\limits_{i=1}^{k} \alpha_i!\right) k^l} \, . \tag{5.18}$$

Similarly,

$$P(\overrightarrow{\beta} \in \mathbb{G}_m \mid m) = \frac{m!}{\left(\prod\limits_{i=1}^{k} \beta_i!\right) k^m} \, . \tag{5.19}$$

Substituting (5.18) and (5.19) into (5.17) we get

$$P(\overrightarrow{\alpha} \in \mathbb{F}_l, \overrightarrow{\beta} \in \mathbb{G}_m \mid l, m) = \frac{l! \, m!}{\left(\prod\limits_{i=1}^{k} \alpha_i! \, \beta_i!\right) k^{l+m}} \, . \tag{5.20}$$

For each of these interference patterns there is a group of associated conditional transition probabilities

$$P(j \mid i, \overrightarrow{\alpha} \in \mathbb{F}_l, \overrightarrow{\beta} \in \mathbb{G}_m) \tag{5.21}$$

that specifies the probability of receiving (detecting) value 'j' given that value 'i' was transmitted given that we have interference patterns $\overrightarrow{\alpha}(l)$ and $\overrightarrow{\beta}(m)$ on the channel. Using the aforementioned arguments, we can deduce that a general expression for the probability of error is given by

$$
P_e = \sum_{\mathbb{E}} P(j \mid i) = \sum_{\mathbb{E}} \sum_{l=0}^{N_u-1} \sum_{m=0}^{N_u-1-l} P(l,m) \overbrace{\sum_{\vec{\alpha} \in \mathbb{F}_l} \sum_{\vec{\beta} \in \mathbb{G}_m} \underbrace{ \begin{matrix} P(\vec{\alpha} \in \mathbb{F}_l,\ \vec{\beta} \in \mathbb{G}_m \mid l,m) \\ \times P(j \mid i,\ \vec{\alpha} \in \mathbb{F}_l,\ \vec{\beta} \in \mathbb{G}_m) \end{matrix} }_{P(j|i\ |\ l,m)} }^{P(j|i)} .
$$

$$(5.22)$$

The terms in (5.22) are defined in (5.10) and (5.20). The only term that is not defined

mathematically is the term given by (5.21). In fact, this term cannot be defined in a

general manner since it is completely dependent on the methods used for modulation,

detection and chip conflict resolution described in Section 5.2.

Up until here, the assumptions made on the OCDMA system are that it is using time

spreading with OOC spreading codes and its receiver is a correlation based receiver. The

fact that we have built our model for a transmitter that has only three states can be easily

relaxed to extend the model to any M-ary system using exactly the same methodology

provided in this section. Hence, the framework provided in this chapter can be used to

deduce the BER for any OCDMA system that fits the aforementioned assumptions.

5.4.2   BER for Absolutely Timed Detection with '1'-Value

First

As an application to our derived generalized BER (P_e) expression we consider the case

of absolutely timed detection and '1'-value first. The absolutely timed detection with

matched filter ODLC decoder is equivalent exactly to the optimum matched filter receiver

described in Chapter 2. This means that the detection variable is the sum of photon

counts over all the location of the pulses within a one bit duration spreading code. On

the other hand, the use of the '1'-value first chip conflict resolution method means that

any '0' and '1' chip interference will be detected as a '1' chip. This means that a pulse

with value '1' sent by the transmitter can never be detected in error as a '0' at the

receiver; mathematically we have

$$P(\text{'0'} \mid \text{'1'}) = 0 \ ,$$

which produces

$$\mathbb{E} = \{(\text{'0'}, \text{'1'}), (\text{'IL'}, \text{'0'}), (\text{'IL'}, \text{'1'})\} \ . \tag{5.23}$$

This means that we need to calculate only $P(j \mid i)$ where $(i, j) \in \mathbb{E}$ as defined in

(5.23). We will start by deducing an expression for (5.21) for different values of $(i, j) \in \mathbb{E}$

then use that to find $P(j \mid i)$, hence, the BER for the scheme.

Starting with the case $(i,j) = ($'0', '1'$)$, a '0' sent by the transmitter can be detected as a '1' by the receiver *if and only if* the number of '1'-interferers is greater than or equal to the threshold $(l \geq h)$ regardless of the number of '0'-interferers. Consequently,

$$P(\text{'1'} \mid \text{'0'}, \overrightarrow{\alpha} \in \mathbb{F}_l, \overrightarrow{\beta} \in \mathbb{G}_m) = P(\text{'0'}) \begin{cases} 1 & l \geq h \\ \\ 0 & \text{elsewhere} \end{cases}, \quad (5.24)$$

where $P(\text{'0'}) = P_A(1 - P_1)$ is the probability of intended source active and sending a '0' bit. Similarly, for the case $(i,j) = ($'IL', '1'$)$ we have

$$P(\text{'1'} \mid \text{'IL'}, \overrightarrow{\alpha} \in \mathbb{F}_l, \overrightarrow{\beta} \in \mathbb{G}_m) = P(\text{'IL'}) \begin{cases} 1 & l \geq h \\ \\ 0 & \text{elsewhere} \end{cases}, \quad (5.25)$$

where $P(\text{'IL'}) = (1 - P_A)$ is the probability that the intended source is idle.

Examining the case $(i,j) = ($'IL', '0'$)$, we find that an idle 'IL' source can be detected as sending a '0' by the receiver *if and only if* $(m \geq h$ and $l < h)$. Hence,

$$P(\text{'0'} \mid \text{'IL'}, \overrightarrow{\alpha} \in \mathbb{F}_l, \overrightarrow{\beta} \in \mathbb{G}_m) = P(\text{'IL'}) \begin{cases} 1 & m \geq h, l < h \\ \\ 0 & \text{elsewhere.} \end{cases} \quad (5.26)$$

Substituting (5.24), (5.25) and (5.26) into (5.22) results in

$$P_e = (1 - P_A P_1) \sum_{l=h}^{N_u-1} \sum_{m=0}^{N_u-1-l} P(l,m) + (1 - P_A) \sum_{l=0}^{h-1} \sum_{m=h}^{N_u-1-l} P(l,m)\phi(l,m) , \qquad (5.27)$$

where

$$\phi(l,m) = \begin{cases} 1 & l+m \le N_u - 1 \\ \\ 0 & \text{elsewhere.} \end{cases}$$

is used to account for the fact that the summation in the second term does not contribute any values for choices of l and m that violate the condition $l+m \le N_u - 1$. With some algebraic manipulations (as shown in Appendix B) (5.27) can be reduced to (5.28) which provides an expression for the probability of error for CLM-BPPM OCDMA system with absolutely timed detection using '1'-value first.

$$P_e = 1 - P_A P_1 + \sum_{l=0}^{h-1} P(l) \left[(1 - P_A)\phi(l) - (1 - P_A P_1) \right] - (1 - P_A) \left(\sum_{l=0}^{h-1} \sum_{m=0}^{h-1} P(l,m)\phi(l) \right) ,$$

$$(5.28)$$

where

$$P(l) = \binom{N_u - 1}{l} q_1^l (1 - q_1)^{N_u-1-l}$$

and

$$\phi(l) = \begin{cases} 1 & l \leq N_u - 1 - h \\ \\ 0 & \text{elsewhere.} \end{cases}$$

5.4.3   Optimal Choice of Parameters

As discussed in Chapters 3 and 4, there is a tradeoff between the parameters affecting the BER of the system. In particular, the choice of the threshold h and the maximum number of multiplexed users (number of orthogonal codewords) N affects both the system capacity as well as the BER. The optimum value for the BER will be a zero value, which is achieved if and only if

$$N \leq h \leq k \ . \tag{5.29}$$

However, fulfilling this condition reduces the system capacity in terms of the number of multiplexed users to a great extent. Another alternative is to aim at maximizing the system capacity by choosing the value of N to be the maximum value supported by the given code design defined by k and n. This value is achieved by choosing N according to

$$N = \left\lfloor \frac{n-1}{k(k-1)} \right\rfloor \ , \tag{5.30}$$

which (under the assumption that n is chosen such that $k(k-1)$ divides $n-1$ as shown in Chapter 6) results in an optimum code length n_{opt} given by

$$n_{opt} = Nk(k-1) + 1 . \tag{5.31}$$

Given this as a system constraint, we can maximize the system performance by choosing the value of the threshold h to be the optimum threshold h_{opt} given by

$$h = h_{opt} = k , \tag{5.32}$$

which minimizes the BER for an OCDMA using OOC and optimal correlator receiver [25].

5.5   Numerical Results

The different parameters that affect our system performance are the code length n, the code weight k, the number of users N_u, the optical threshold h, the probability of activity per source P_A, and the probability of an active source sending a '1' bit P_1. In our results, we assume that $P_1 = 0.5$ and the system is working with optimal threshold value $h = k$ unless otherwise stated. We use simulation as well as the analytical model provided earlier in this chapter in order to derive three main sets of results. In the first set of results we compare the analytical model with simulations in order to verify both under different system parameters. We then provide a more comprehensive set of results that shows how the different system parameters affect the BER performance. The last set of results focuses on the operation of the system under the optimal parameter values that minimizes the probability of error subject to using the maximum number of multiplexed users as stated in (5.31) and (5.32).

It can be seen from Figure 5.7 that our analytical results closely follow the results obtained from simulations for different system parameters. This verifies the accuracy of our analytical approach.

Figures 5.8 and 5.9 show that increasing the code length n decreases the BER P_e, which is expected because longer spreading codes with the same code weight means lower probability of pulse overlap, which means lower probability of error. It can be also deduced from these two figures that increasing the number of users N_u results in higher probability of error due to the increase in the average number of pulses overlapping at a single chip spot. This means a higher probability for an overlap between number of pulses greater than or equal to the optical threshold h, which increases the probability of detection errors.

One thing that is worth noting in Figures 5.8 and 5.9 is that increasing the code weight k results in decreasing P_e. This phenomena, which is verified in Figure 5.10, contradicts the intuition that suggests that increasing the number of pulses per code (code weight k) should result in higher probability of pulse overlaps, which in turn will result in increasing P_e as can be deduced from (5.27). However, if we notice that for an optimum system operation, we always set the optical threshold h to be equal to the code weight k, we deduce that it is this assumption that produces the previously mentioned effect. In other words, an increase in the code weight is always coupled with an equal increase in the optical detection threshold h. What the results actually show is that an increase in h has more dominant effect on reducing the P_e than the increase caused by increasing k. The decoupling between the two parameters is demonstrated in Figure 5.11.

Figure 5.11 shows how increasing the optical threshold h results in a significant enhancement in the system BER. It must be noted, however, that h is never allowed to be higher than the code weight k in order for the system to be working properly. The figure also shows that if we fix the value of h and increase k, P_e in fact increases as explained in the previous paragraph.

In Figure 5.12, we investigate the effect of the source (user) average activity represented by the probability of source being active P_A on the BER. It can be deduced from the figure that increasing the average source activity results in higher BER. Again, this is expected, because an increase in the average source activity means higher number of users sending '1' and '0' data bits, which means higher probability of overlapped pulses. The same figure also shows that generally speaking OOK slightly outperforms BPPM in terms of BER for the same system parameters. Both techniques provide identical BER for the case of fully loaded sources ($P_A = 1$). However, OOK provides a slightly better BER, BPPM has the advantages of easier synchronization (clock recovery) and having the ability to distinguish between an idle and an active source.

Finally, Figures 5.13 and 5.14 show the behavior of the system under the maximum

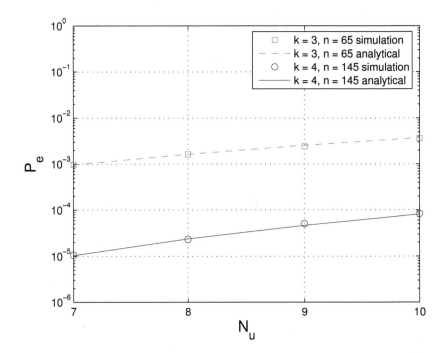

Figure 5.7: Comparison of BER from mathematical analysis and simulation.

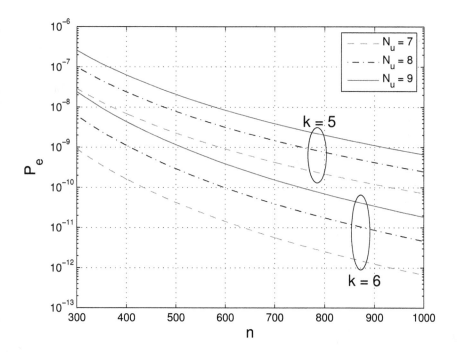

Figure 5.8: Effect of changing code length on BER.

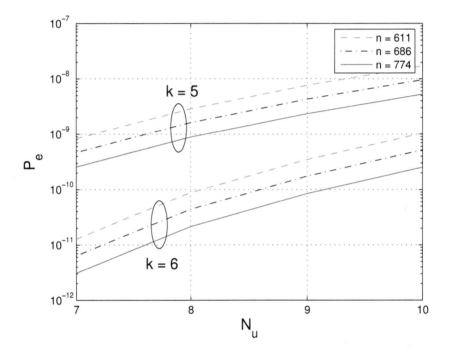

Figure 5.9: Effect of increasing number of users on BER.

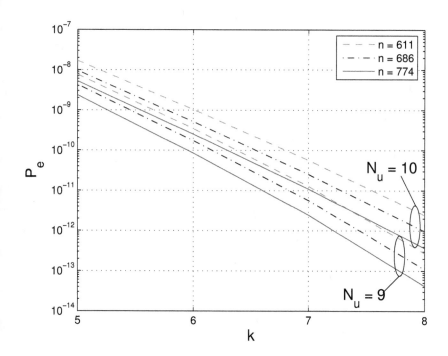

Figure 5.10: Effect of changing code weight on BER.

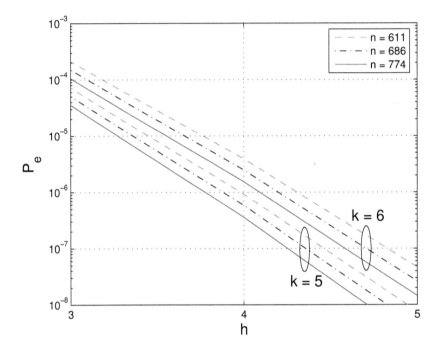

Figure 5.11: Effect of changing the optical threshold on BER ($N_u = 9$).

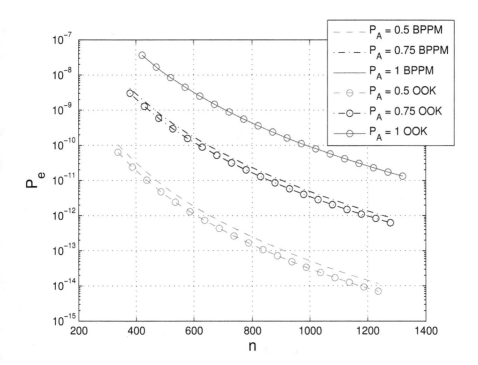

Figure 5.12: Effect of increasing average user activity P_A on BER ($k = 7$).

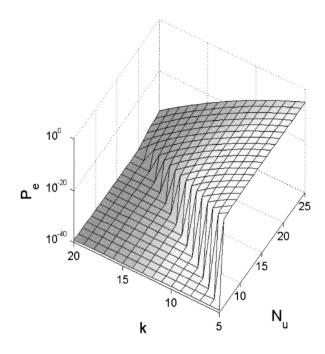

Figure 5.13: BER for a system using OOC with optimum code length.

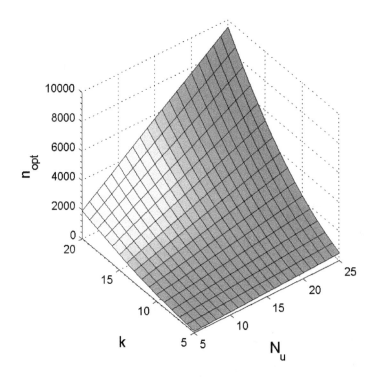

Figure 5.14: Optimum code length as a function of code weight and number of users.

capacity condition given by (5.31) and minimum BER condition given by (5.32). It can

be seen from Figure 5.13 that the system BER has two regions; one with zero MAI errors,

which is the region defined by the condition $N_u < h_{opt}$ and the other is a region with

MAI errors. In the region with MAI errors the BER of the system increases with N_u and

decreases with k due to the associated increase in h. It must be noted, however, that the

effect of the coupled change in k and h on the BER is more prominent than the effect of

N_u on the BER. This is demonstrated by the rapid non-linear change in P_e with k and

h as compared to the slower close to linear change in P_e with N_u shown in the figure.

Figure 5.14 shows the value of n_{opt} associated with each value of P_e in Figure 5.13.

5.6   Concluding Remarks

We have proposed a modulation method for optical fiber CDMA transmission based

on encoding the bit level value at the chip level of the spread signal using binary pulse

position modulation, hence, the name Chip-Level Modulated BPPM (CLM-BPPM). Our

modulation method maintains the advantages of PPM modulation over OOK methods by

allowing for easier clock recovery at the receiver side as well as an ability to detect source

activity. Moreover, our method surpasses the previously proposed PPM methods based

on bit level modulation because it preserves the autocorrelation and crosscorrelation

properties of the optical spreading codes across the entire bit duration. An architecture

for an optical communication system that employs the proposed scheme is provided. We

have provide several variations of the main scheme based on applying different methods for detection and bit conflict resolution.

In order to evaluate the performance of our proposed scheme based on bit error rate (BER) due to multiple access interference (MAI), we started by developing a general mathematical framework to describe the BER in terms of different system parameters. Our mathematical model incorporated for the first time the average source activity statistics into the BER calculations.

We have applied the developed framework on one of the variations of our proposed scheme to get an expression for the BER. Using our developed mathematical model we have numerically evaluated the BER of CLM-BPPM with absolutely timed detection and '1'-value first chip conflict resolution and compared the results with simulation results to verify the mathematical model. The comparison shows that our mathematical model accurately represents the system performance under the given assumptions.

From the results, we have seen that increasing the number of users increases the BER due to the higher level of interference present in the system. Our results also show that if the optimality conditions are maintained, increasing the code weight enhances the system performance due to the fact that it requires increasing the optical detection threshold, which means higher ability to suppress MAI. Separating the changes of code weight and optical detection threshold, we notice that in fact increasing the code weight alone while fixing the threshold results in increasing the BER.

We have also examined the effect of the average source activity on the BER and have shown that increasing the average source activity increases the BER of the system.

Compared to an OOK system, our CLM-BPPM system has a slightly higher BER, but as the source activity increases (load is increased), both systems show closer BER values. Asymptotically, a fully loaded system ($P_A = 1$) produces the same BER for OOK and CLM-BPPM.

The results also demonstrate that the BER of a CLM-BPPM system operating at the optimum BER point ($h = k$) is highly sensitive to changes in the code weight k, which suggests that network designers should focus more on choosing the optimum value for the spreading code weight.

In this chapter, we have developed the CLM-BPPM modulation scheme for OCDMA systems. We have also developed a general mathematical framework for calculating the BER for time-spread OCDMA systems. The framework was used to deduce the BER of our CLM-BPPM modulation technique.

In the next chapter, we focus on developing efficient construction methods for OOC spreading codes. Our aim is to develop a low complexity algorithm that can construct bandwidth-efficient spreading codes.

Chapter 6

OOC Construction Using Rejected

Delays Reuse

In Chapter 5, we have proposed CLM-BPPM as a modulation method for OCDMA
systems. We have also developed a general mathematical framework for the BER of
time-spread OCDMA systems and used our developed model to deduce the BER for
CLM-BPPM. In this chapter, we shift our focus to the problem of optical spreading
code design. In particular, we focus on the design of OOC using greedy algorithms. We
introduce a modified element-by-element greedy algorithm for constructing OOC with
small code length given a certain code weight and number of codewords. Our modified
algorithm is called Greedy Algorithm with Rejected Delays Reuse (RDR).

The rest of this chapter is organized as follows. Section 6.1 is a brief introduction. In
Section 6.2, we provide mathematical definitions and representations for OOCs and their
main properties. Following that, in Section 6.3, an overview of construction methods for
OOCs as well as a detailed description of a computationally efficient element-by-element
greedy algorithm for constructing OOCs are provided. In Section 6.4, we provide a theorem that shows how the greedy algorithm provided in Section 6.3 can be enhanced then

we introduce our new RDR construction method. In Section 6.5, we evaluate the efficiency of OOCs generated using RDR as compared to those generated using the existing element-by-element greedy algorithm by introducing an efficiency factor that provides a quantitative measure of the code's ability to expand sub-wavelength switching capacity. We call this factor the expansion efficiency factor. In Section 6.6, we provide numerical results that show how our newly designed OOCs outperform those designed using the original element-by-element greedy construction. Finally, we conclude the chapter in Section 6.7.

6.1 Introduction

The use of OCDMA as a multiplexing (switching) layer at the sub-wavelength level in Generalized Multi-Protocol Label Switching (GMPLS) networks was introduced in Chapter 3 by proposing the OC-GMPLS architecture. In order to use OCDMA, the optical signal is spread either in the time domain (temporal spreading) using Fiber Tapped Delay Lines or in the wavelength domain (spectral spreading) using Arrayed Waveguide Grating (AWG) or Fiber Bragg Grating (FBG). In both cases a spreading code is required. Optical Orthogonal Codes (OOC) [66] are widely used as optical spreading codes because they achieve the optimum criteria for unipolar $\{0,1\}$ codes in terms of their autocorrelation and crosscorrelation properties. However, efficient OOCs are difficult to construct.

In [85], Argon *et al* presented an algorithm for constructing OOCs based on the extended set representation. Being a member of the class of greedy algorithms, the algorithm did not reconsider previously rejected delay elements in future construction steps. In this chapter, we show that rejected elements at one step can in fact be accepted in later steps. Based on this fact, we propose a new algorithm called Greedy Algorithm with Rejected Delays Reuse (RDR) to construct OOCs. We show through our numerical results that OOCs designed using RDR are generally shorter than those constructed using the classical element-by-element greedy algorithm. In order to study the effect of using shorter codes on switching capacity we define a switching capacity expansion efficiency factor and show that shorter codes lead to higher expansion efficiency factors, which means higher switching capacity.

6.2 Definitions and Mathematical Preliminaries

In the following, we provide some mathematical definitions of Optical Orthogonal Codes and their set theoretic representations. Then, we provide concrete mathematical formulas that can be used to deduce the interdelay set theoretic representation of an OOC from its direct classical definition. Using our mathematical representation, as well as the mathematical properties of OOCs, we provide new proofs for two well known theorems on OOCs.

In Chapter 3, we have stated the conditions on the autocorrelation and crosscorrelation properties of an OOC. In the following, we provide a more mathematically solid definition of optical orthogonal codes and their different properties.

Definition 6.1. *A* $(n, k, \lambda_a, \lambda_c)$ *OOC, $\mathcal{C}$ with cardinality $|\mathcal{C}| = N$, is defined as a family*

of N $\{0,1\}$-sequences of length n and weight k with circular autocorrelation $\mathcal{R}_{XX}(\tau)$

for sequence $X \in \mathcal{C}$ and circular crosscorrelation $\mathcal{R}_{XY}(\tau)$ between any two sequences

$X, Y \in \mathcal{C}$ satisfying the following properties:

$$\mathcal{R}_{XX}(\tau) = \left| \sum_{j=0}^{n-1} x_j \cdot x_{j \oplus \tau} \right| = \begin{cases} k & \tau = 0 \\ \\ \leq \lambda_a & \tau \neq 0 \end{cases} \tag{6.1}$$

$$\mathcal{R}_{XY}(\tau) = \left| \sum_{j=0}^{n-1} x_j \cdot y_{j \oplus \tau} \right| \leq \lambda_c \, , \tag{6.2}$$

where x_j is element number j in code X, y_j is element number j in code Y and $\oplus$

represents addition modulo-n.

In this chapter, we consider the case $\lambda_a = \lambda_c = 1$. We call this a $(n, k, 1)$ OOC and will

refer to it as OOC for simplicity. This is the case with the lowest multiuser interference in

an OCDMA system based on OOCs, which enhances the system performance. To avoid

trivial code constructions, we consider only the case where $N \geq 1$ and $n > k > 1$.

OOCs constructed using the above definitions will constitute sparse matrices. A

$(n, k, 1)$ OOC consists of an $N \times n$ matrix with $N \times k$ '1's (k in each row) and the

remaining elements being '0's, where $k << n$. Thus, a direct representation of an OOC

is not convenient. It is more compact and mathematically tractable to represent an OOC

by its set of interdelays $\mathbb{T}$ between the locations of the '1' bits in each codeword. Another

representation of OOC called the *set theoretic representation* uses sets that contain the column numbers of the '1's in $\mathcal{C}$. In the rest of this chapter, we will use the set of interdelays $\mathbb{T}$ to represent an OOC. Each row of $\mathbb{T}$ represents a codeword from the OOC. The number of rows is the number of codewords and the number of columns is the code weight.

Definition 6.2. *The set of interdelays $\mathbb{T}$ for an OOC $\mathcal{C}$ is the set that contains the differences in column numbers (relative positions) between any two adjacent '1's in each codeword (row) of $\mathcal{C}$, treating each codeword as a circularly connected subset. Formally,*

$$\mathbb{T} = \left\{ t_{i,j} : t_{i,j} = \begin{cases} col(1_{i,j+1}) - col(1_{i,j}) \ , \ 1 \le j \le k-1 \\ \\ col(1_{i,1}) + n - col(1_{i,k}) \ , \ j = k \end{cases} \ , \ 1 \le i \le N \right\},$$

where $1_{i,j}$ is the j_{th} '1' bit in row (codeword) number i, $t_{i,j}$ is element in row i and column j of $\mathbb{T}$, and $col(c_{i,j}) = j$ is the column number of element $c_{i,j} \in \mathcal{C}$.

Definition 6.3. *The extended set of interdelays $\mathbb{T}_{ext}$ for an OOC $\mathcal{C}$ is the adjoint of the set $\mathbb{T}$ with the set that contains the summations of adjacent interdelays, up to $(k-1)$ adjacent interdelays, in each row of $\mathbb{T}$, treating each row as a circularly connected subset.*

Formally,

$$\mathbb{T}_{ext} = \{e_{i,j} : e_{i,j} = \sum_{p=q_j}^{q_j+u_j} t_{i,v_p},$$

$$1 \leq i \leq N, \quad 1 \leq j \leq k(k-1),$$

$$q_j = \left\lfloor \frac{j-1}{k-1} \right\rfloor + 1, \quad u_j = (j-1) \bmod (k-1),$$

$$v_p = (p-1 \bmod k) + 1\}.$$

In order for a code to satisfy inequalities (6.1) and (6.2), some conditions have to be satisfied by $\mathbb{T}_{ext}$, as stated by the following theorem.

Theorem 6.4. *For a set of interdelays $\mathbb{T}$ to represent an OOC $\mathcal{C}$, it must satisfy two conditions:*

1. *The set of interdelays $\mathbb{T}$ must have no repeated elements, i.e.,*

$$t_{i,j} \neq t_{l,m} \forall i,j,l,m \text{ such that } i \neq l \ , j \neq m.$$

2. *The extended set of interdelays $\mathbb{T}_{ext}$ must have no repeated elements, i.e.,*

$$e_{i,j} \neq e_{l,m} \forall i,j,l,m \text{ such that } i \neq l \ , j \neq m.$$

Proof. Since, by definition $\mathbb{T} \subset \mathbb{T}_{ext}$, it suffices to prove part 2 of the theorem. Part 1 will follow implicitly.

From the definition of the autocorrelation and crosscorrelation functions, given in (6.1) and (6.2) and knowing that $x_i \in \{0, 1\}$ and $y_i \in \{0, 1\}$, it follows that $\mathcal{R}_{XX}(\tau)$ represents the summation of overlapping '1' bits between a codeword X and its τ-bits rotated version. Also, $\mathcal{R}_{XY}(\tau)$ is the summation of overlapping '1' bits between a codeword X and a τ-bits rotated version of another codeword Y.

For an OOC, we require that $\lambda_a = \lambda_c = 1$. This means that for any value of $\tau \neq 0$ we require that there will be a maximum of a single '1' bit overlap between a codeword and its τ-bits rotated version. Similarly, for any value of τ there will be at most a single '1' bit overlap between a codeword and a τ-bits rotated version of another codeword.

Taking the case of autocorrelation and denoting the τ-bits rotated version of X by $\hat{X}$, we see that for the case $\tau = 0$ all the '1' bits from codeword X will overlap with the '1' bits from the 0-bits rotated version $\hat{X}$. Since, we have k '1' bits, we get $\mathcal{R}_{XX}(0) = k$, which satisfies the condition for $\mathcal{R}_{XX}(0)$. If $\tau \neq 0$, without loss of generality, we can always rotate X and $\hat{X}$ so that an arbitrary '1' bit from X overlaps with an arbitrary '1' bit from $\hat{X}$ at location $\tau_1 = 1$. Another '1' bit from the sequence X at location $\tau_1 + \delta\tau_\alpha$ overlaps with a '1' bit from the sequence $\hat{X}$ at location $\tau_1 + \delta\tau_\beta$ *if and only if* $\delta\tau_\alpha = \delta\tau_\beta$, where $\delta\tau_m$ indicates the distance between the m_{th} '1' bit and the location of the first overlap τ_1. Accordingly, a *sufficient and necessary condition* for a single overlap is that

$\delta\tau_\beta \neq \delta\tau_\alpha$ for all values of $\alpha \in [2,k]$ and $\beta \in [2,k]$. However,

$$\delta\tau_m = \sum_{j=2}^{m} t_{i,j},$$

which represents the elements $e_{i,j} \in \mathbb{T}_{ext}$ taking into consideration the arbitrary rotation performed to make the first '1' bit overlap at location τ_1 of the codeword. Therefore, the *sufficient and necessary condition* for a codeword X at row i of $\mathcal{C}$ to satisfy the autocorrelation constraint is

$$e_{i,j} \neq e_{i,m} \forall i, j, m \text{ such that } j \neq m.$$

Using the same method it can be shown that the *sufficient and necessary condition* for two codewords X and Y at rows i and l of $\mathcal{C}$ to satisfy the crosscorrelation constraint is

$$e_{i,j} \neq e_{l,m} \forall j, m \text{ such that } j \neq m \forall i \neq l.$$

Combining both conditions we get

$$e_{i,j} \neq e_{l,m} \forall i, j, l, m \text{ such that } i \neq l \text{ , } j \neq m,$$

and the theorem follows. $\square$

For a code that satisfies Theorem 6.4, a Johnson bound [16] can be found on its cardinality. This bound is given in the following theorem.

Theorem 6.5. *For a $(n, k, 1)$ OOC, $\mathcal{C}$, with cardinality $|\mathcal{C}| = N$, the following inequality always holds*

$$N \leq \left\lfloor \frac{n-1}{k(k-1)} \right\rfloor$$

Proof. Using the set of interdelays $\mathbb{T}$ to represent $\mathcal{C}$, for every codeword constructed there is a corresponding row $\mathbb{T}_{ext}(r)$ in the extended set $\mathbb{T}_{ext}$, where r is the row number, with cardinality $|\mathbb{T}_{ext}(r)| = k(k-1)$. From Theorem 6.4, none of the elements $\{e_{r,j} \in \mathbb{T}_{ext}(r)\}$ can repeat in other extended subsets (rows) $\{e_{i,j} \in \mathbb{T}_{ext}(i) : i \neq r\}$ for other codewords in $\mathcal{C}$. Hence, every codeword consists of unique $k(k-1)$ elements from $\mathbb{T}_{ext}$. For a *non-trivial OOC* $\mathcal{C}$ with length $n > 1$ and weight $k > 1$, we have $1 \leq e_{i,j} \leq n-1$. Accordingly, we have a pool of $n-1$ elements (integer numbers) out of which we consume $k(k-1)$ elements per constructed codeword. This means that the maximum number of constructed codewords N is less than or equal to $\frac{n-1}{k(k-1)}$. Since this number must be an integer number, we have

$$N \leq \left\lfloor \frac{n-1}{k(k-1)} \right\rfloor$$

$\square$

6.3 Classical Element-by-Element Greedy Method

for Constructing OOCs

Designing an OOC can be formulated as an optimization problem. An OOC design is the construction of a set of N $\{0,1\}$-sequences with equal lengths n, and equal weights k that satisfy the autocorrelation and crosscorrelation properties stated in (6.1) and (6.2) with $\lambda_a = \lambda_c = 1$, subject to minimizing n for a given k and N.

The methods of combinatorial designs are the roots of many code design methods for OOC [86,87]. An OOC can be constructed using known combinatorial methods (designs) by appropriately mapping the OOC design problem into one of the solved combinatorial design problems. Then the solution can be used to design the OOC. In [16], Chung *et al* proposed the *greedy algorithm* for constructing the code set $\mathcal{C}$ as well as several other techniques. The proposed greedy algorithm incorporates an exhaustive search through all the possible $\binom{n}{k}$ codewords in order to find the codewords that satisfy (6.1) and (6.2). These codewords constitute $\mathcal{C}$. Later in the same paper, the authors proposed an accelerated greedy algorithm, which constructs $\mathcal{C}$ using an iterative method over its set theoretic representation.

In [85], Argon *et al* presented a variation of the accelerated greedy algorithm (Algorithm 6.1). The algorithm reduces the number of construction steps by using an iterative element-by-element greedy construction method for building the interdelays set representation $\mathbb{T}$ of an OOC. The algorithm starts with an empty set and uses two nested loops

---

Algorithm 6.1 OOCGreedy

Require: $N \geq 2, k \geq 2$
Ensure: $\mathcal{R}_{XX} \leq 1, \mathcal{R}_{XY} \leq 1 \forall X, Y \in \mathcal{C}$ such that $X \neq Y$

1: $e \leftarrow 1$
2: **for all** i such that $1 \leq i \leq N$ **do**
3:    $\mathbb{T}(i, 1) \leftarrow e$
4:    $e \leftarrow e + 1$
5: **end for**
6: **for** $j = 2$ to $k - 1$ **do**
7:    **for** $i = 1$ to N **do**
8:       $O \leftarrow FALSE$
9:       **while** $\neg O$ **do**
10:          $\mathbb{T}(i, j) \leftarrow e$
11:          $O \leftarrow$ CheckIntermediateOrthogonal($\mathbb{T}$)
12:          $e \leftarrow e + 1$
13:       **end while**
14:    **end for**
15: **end for**
16: $O \leftarrow FALSE$
17: **for all** i such that $1 \leq i \leq N$ **do**
18:    $m_x(i) \leftarrow \sum\limits_{j=1}^{k-1} \mathbb{T}(i, j)$
19: **end for**
20: $M_x \leftarrow \max\limits_{i}\{m_x(i)\}$
21: **while** $\neg O$ **do**
22:    **for all** i such that $1 \leq i \leq N$ **do**
23:       $\mathbb{T}(i, k) \leftarrow e + (M_x - m_x(i))$
24:    **end for**
25:    $O \leftarrow$ CheckCompleteOrthogonal($\mathbb{T}$)
26:    $e \leftarrow e + 1$
27: **end while**

---

to construct the set of interdelays representation of the designed OOC. At step (i, j), the outer loop points at codeword number i. All the codewords at rows $\{1, ..., i-1\}$ are of length j, while the remaining codewords are of length $j - 1$. The inner loop tries to find the smallest possible untested interdelay element d_j at column number j to add to codeword (row) number i such that orthogonality conditions are satisfied.

The functions *CheckIntermediateOrthogonal*(S) and *CheckCompeleteOrthogonal*(S) perform the orthogonality check (i.e., ensure that Theorem 6.4 is satisfied) of the OOC under construction. These functions are implemented by Algorithms 6.2 and 6.3 respectively. The major two steps in these functions are:

1. construct the extended set $\mathbb{T}_{ext}$

2. ensure that there are no repeated elements in $\mathbb{T}_{ext}$

The difference between the first function (Algorithm 6.2) and the second function (Algorithm 6.3) is in the first step (i.e., the construction of $\mathbb{T}_{ext}$). The check for repeated elements in a set can be easily achieved using a quick sort followed by a linear check of every element and its immediate following neighbor. If there are no matches, then there are no repeated elements.

Algorithm 6.2 CheckIntermediateOrthogonal(S)

Require: $\mathbb{T} \neq \phi$
Ensure: $\mathbb{T}_{ext}$ has no repeated elements
 1: $\mathbb{T}_{ext} \leftarrow \mathbb{T}$
 2: $j_{max} \leftarrow k$
 3: $ind \leftarrow k + 1$
 4: **for all** i such that $1 \leq i \leq N$ **do**
 5:     **for** $p = 1$ to $k - 1$ **do**
 6:         **for** $j = p + 1$ to j_{max} **do**
 7:             $\mathbb{T}_{ext}(i, ind) \leftarrow \sum\limits_{l=p}^{l=j} \mathbb{T}(i, l)$
 8:             $ind \leftarrow ind + 1$
 9:         **end for**
10:     **end for**
11: **end for**
12: **if** $\mathbb{T}_{ext}$ has repeated elements **then**
13:     $O \leftarrow$ FALSE
14: **else**
15:     $O \leftarrow$ TRUE
16: **end if**
17: **return** O

Algorithm 6.3 CheckCompleteOrthogonal(S)

Require: $\mathbb{T} \neq \phi$

Ensure: $\mathbb{T}_{ext}$ has no repeated elements

1: $\mathbb{T}_{ext} \leftarrow \mathbb{T}$
2: $ind \leftarrow k + 1$
3: **for all** i such that $1 \leq i \leq N$ **do**
4:     **for** $p = 1$ to $k - 1$ **do**
5:        $j_{max} \leftarrow p + k - 2$
6:        **for** $j = p + 1$ to j_{max} **do**
7:           **if** $j \leq k$ **then**
8:              $\mathbb{T}_{ext}(i, ind) \leftarrow \sum_{l=p}^{l=j} \mathbb{T}(i, l)$
9:           **else**
10:             $\mathbb{T}_{ext}(i, ind) \leftarrow \sum_{l=p}^{l=k} \mathbb{T}(i, l) + \sum_{l=1}^{l=j-k} \mathbb{T}(i, l)$
11:           **end if**
12:           $ind \leftarrow ind + 1$
13:        **end for**
14:     **end for**
15: **end for**
16: **if** $\mathbb{T}_{ext}$ has repeated elements **then**
17:     $O \leftarrow$ FALSE
18: **else**
19:     $O \leftarrow$ TRUE
20: **end if**
21: **return** O

6.4 Proposed Greedy Algorithm with Rejected

Delays Reuse (RDR)

Algorithm 6.1 follows the classical greedy algorithm argument, which mandates that an interdelay value that has been tested in a previous step and found to cause loss of orthogonality should never be revisited in later steps of the solution. This argument might look intuitively correct. However, in the following we present a theorem that shows that a retest of such rejected elements might succeed in using them later on.

Theorem 6.6. *An interdelay value $d_{i,j}$ that has been tested and failed to maintain orthogonality in step (i,j) (i.e., working on element j of codeword i) of the greedy construction for OOC C given in Algorithm 6.1 can possibly be reused at another step (l,m) such that the pair (l,m) is distinct from the pair (i,j).*

Proof. There are two scenarios for rejecting an interdelay element $d_{i,j}$ in step (i,j):

- The first scenario is when before including $d_{i,j}$ in $\mathbb{T}$,

$$\exists e_{i,j} \in \mathbb{T}_{ext} : e_{i,j} = d_{i,j}.$$

In this case, it is obvious that the inclusion of $d_{i,j}$ did not affect the rejection condition, hence, $d_{i,j}$ can never be included in any later steps.

- The second scenario is when after including $d_{i,j}$ in $\mathbb{T}$,

$$\exists e_{i,j} \, , \, \exists e_{l,m} \in \mathbb{T}_{ext} : e_{i,j} = e_{l,m} \text{ and } i \neq l, j \neq m$$

with $d_{i,j}$ used to either construct $e_{i,j}$ or $e_{l,m}$

In this case the rejection condition is dependent not only on the value of $d_{i,j}$, but also on the exact step in which it is included in $\mathbb{T}$.

The second case happens because the calculation of the extended set elements depends on the adjacency between interdelay values in $\mathbb{T}$. Consequently, by delaying the inclusion of $d_{i,j}$ to a later step, we might be able to avoid the rejection condition. □

Another way to prove Theorem 6.6 is by giving a single example that fulfills what the theorem states, which we provide in Figure 6.1 later in this section.

Theorem 6.6 suggests that a retest of a rejected element might result in using it while constructing an OOC. Since a rejected element in the greedy algorithm will always be smaller than any other elements tested in later steps, a reuse of a rejected element will *locally minimize the code length n*, which is desirable in the greedy algorithm in hope that the final solution will result in a shorter code length. Based on this argument, we have developed a modified greedy algorithm called *Rejected Delays Reuse (RDR) Greedy Algorithm* (Algorithm 6.4 below).

The algorithm is similar to the classical algorithm (Algorithm 6.1), but with a major modification that allows for reusing a previously rejected delay element if possible. The

---

Algorithm 6.4 OOCRDRGreedy

Require: $N \geq 2, k \geq 2$

Ensure: $\mathcal{R}_{XX} \leq 1, \mathcal{R}_{XY} \leq 1 \forall X, Y \in \mathcal{C}$ such that $X \neq Y$

1:   $e \leftarrow 1$, $\mathbb{RD} \leftarrow \phi$

2:   **for all** i such that $1 \leq i \leq N$ **do**

3:      $\mathbb{T}(i,1) \leftarrow e$, $e \leftarrow e + 1$

4:   **end for**

5:   **for** $j = 2$ to $k - 1$ **do**

6:      **for** $i = 1$ to N **do**

7:         $O \leftarrow FALSE$

8:         **for** $l = 1$ to $|\mathbb{RD}|$ **do**

9:             $\mathbb{T}(i,j) \leftarrow \mathbb{RD}(l)$, $O \leftarrow$ CheckIntermediateOrthogonal($\mathbb{T}$)

10:           **if** O **then**

11:             $\mathbb{RD} \leftarrow \mathbb{RD} - \{\mathbb{RD}(l)\}$,   BREAK

12:           **end if**

13:         **end for**

14:         **while** $\neg O$ **do**

15:           $\mathbb{T}(i,j) \leftarrow e$, $O \leftarrow$ CheckIntermediateOrthogonal($\mathbb{T}$)

16:           **if** $\neg O$ **then**

17:             $\mathbb{RD} \leftarrow \mathbb{RD} + \{e\}$

18:           **end if**

19:           $e \leftarrow e + 1$

20:         **end while**

21:      **end for**

22: **end for**

23: $O \leftarrow FALSE$

24: **for all** i such that $1 \leq i \leq N$ **do**

25:      $m_x(i) \leftarrow \sum\limits_{j=1}^{k-1} \mathbb{T}(i,j)$

26: **end for**

27: $M_x \leftarrow \max\limits_{i}\{m_x(i)\}$

28: **for** $l = 1$ to $|\mathbb{RD}|$ **do**

29:      **for all** i such that $1 \leq i \leq N$ **do**

30:         $\mathbb{T}(i,k) \leftarrow \mathbb{RD}(l) + (M_x - m_x(i))$

31:      **end for**

32:      $O \leftarrow$ CheckCompleteOrthogonal($\mathbb{T}$)

33:      **if** O **then**

34:         BREAK

35:      **end if**

36: **end for**

37: **while** $\neg O$ **do**

38:      **for all** i such that $1 \leq i \leq N$ **do**

39:         $\mathbb{T}(i,k) \leftarrow e + (M_x - m_x(i))$

40:      **end for**

41:      $O \leftarrow$ CheckCompleteOrthogonal($\mathbb{T}$)   , $e \leftarrow e + 1$

42: **end while**

---

main modifications can be described as follows.

In step (i, j), if element $d_{i,j}$ has been rejected, it is added to a list of rejected elements. When we move to step (l, m) such that $(l > i$ AND $m = j)$ OR $m > j$, before trying a new element, we retest the elements in the rejected list, starting with the smallest in ascending order. If one of these elements can be used, we use it and remove it from the rejected list. Otherwise, we test new elements. This process continues until the last element of $\mathbb{T}$ has been found.

There are some extra modifications that can be added to the algorithm in order to increase its efficiency. One such modification is to restrict the list of rejected elements to elements that correspond to the second scenario in the proof of Theorem 6.6. This would save both space and computation time, although it would complicate the algorithm description making it more difficult to understand.

Since the RDR algorithm is typically used off-line to calculate the OOC once and then use the same code for OCDMA communications, calculation efficiency is not a major factor at this point and we will sacrifice the efficiency for the sake of clarity. The RDR algorithm can be implemented asymptotically in $O(n^2)$ computation time, as proven in Appendix C. This computation time can be achieved in a fraction of a second on a GHz processor for codes of lengths in the order of $n \sim 1000$. This is considerably faster than the classical element-by-element greedy method given in [85] taking into consideration the fact that RDR codes produce smaller values for n while performing approximately the same number of steps.

We observed while running the algorithm that the terminating elements of codewords (i.e., elements at column number k) are always much larger than all the other interdelay elements of the codeword. This suggests that testing the rejected elements at the finalization step can be omitted to save computation time without affecting the resulting codes. The larger values of the terminating elements are due to the fact that orthogonality conditions applied to these elements are more strict since they include circular adjacency into the calculation as shown in Algorithm 6.3. Another restriction on the values of these elements is that they have to be selected such that all the codewords have the same length n as shown in Algorithms 6.1 and 6.4. If this condition is relaxed (i.e., variable codewords are allowed) it would greatly reduce the magnitude of the terminating interdelays.

OOCs designed using Algorithm 6.4 with their corresponding (i.e., having the same parameters) codes derived using Algorithm 6.1 are shown in Figure 6.1. One can notice that once an element is reused by the RDR algorithm (e.g., the value 27 at the *4th* location in the last codeword of the $\{7, 4\}$ code), the elements after that are generally smaller than those used for the classical greedy algorithm, which results in a code with smaller length.

$\{k, N\}$	$\mathbb{T}$	$\mathbb{T}_{RDR}$
$\{7, 4\}$	$\{1, 5, 12, 24, 31, 50, 144\}$	$\{1, 5, 12, 24, 16, 48, 139\}$
	$\{2, 7, 13, 25, 32, 58, 130\}$	$\{2, 7, 13, 25, 30, 49, 119\}$
	$\{3, 8, 15, 28, 37, 60, 116\}$	$\{3, 8, 15, 28, 31, 53, 107\}$
	$\{4, 10, 19, 30, 44, 62, 98\}$	$\{4, 10, 19, 27, 34, 35, 116\}$
$\{6, 6\}$	$\{1, 7, 17, 28, 50, 174\}$	$\{1, 7, 17, 28, 44, 150\}$
	$\{2, 9, 18, 31, 51, 166\}$	$\{2, 9, 18, 31, 46, 141\}$
	$\{3, 10, 20, 35, 57, 152\}$	$\{3, 10, 20, 35, 48, 131\}$
	$\{4, 12, 22, 39, 59, 141\}$	$\{4, 12, 22, 32, 39, 138\}$
	$\{5, 14, 23, 43, 62, 130\}$	$\{5, 14, 23, 36, 51, 118\}$
	$\{6, 15, 26, 46, 67, 117\}$	$\{6, 15, 26, 43, 56, 101\}$

Figure 6.1: Examples of OOC designs using RDR.

6.5 Sub-Wavelength Switching Capacity and Code Efficiency Analysis

For an optical multiplexing system based on Code Division Multiple Access (OCDMA) at the sub-wavelength level [25], it is desired to be able to multiplex as many users as possible per wavelength channel without affecting the system error rate under ideal conditions. It was shown in [25] that the condition for an OCDMA system to operate as an error free transmission system under ideal conditions is to be able to completely suppress the Multiple Access Interference (MAI) caused by unwanted users on the intended received signal using an optical threshold (hard limiter) device at the output of the correlator receiver. This is possible if and only if $N \leq k$.

In order to maximize the system capacity (maximize the number of multiplexed users per wavelength) we take

$$N = k. \tag{6.3}$$

It was also shown in [25] that the label space expansion ratio ρ, which measures the increase in the system capacity per wavelength of a sub-wavelength OCDMA switched system as compared to a pure Wavelength Division Multiplexing (WDM) system, is given by:

$$\rho = N. \tag{6.4}$$

Using (6.4), ρ for an error-free system is given by:

$$\rho = k. \tag{6.5}$$

From Theorem 6.5, we get

$$N \leq \left\lfloor \frac{n-1}{k(k-1)} \right\rfloor, \tag{6.6}$$

which yields a maximum number of sub-wavelength multiplexed users (maximum system capacity per wavelength) given by:

$$N = \left\lfloor \frac{n-1}{k(k-1)} \right\rfloor, \tag{6.7}$$

which can be rewritten as

$$N \cdot k(k-1) = n - 1 - (n-1 \mod k(k-1)). \tag{6.8}$$

In order to make the maximum use of the available system bandwidth, it is desired to have the smallest n that satisfies (6.8). Inspecting (6.8), the minimum value for n for given values of k and N is achieved when $(n-1 \mod k(k-1)) = 0$. This happens when

$$k(k-1) \mid n-1, \tag{6.9}$$

which reads $k(k-1)$ divides $n-1$. Under this condition, (6.8) can be simplified to

$$n_{min} = N \cdot k(k-1) + 1, \tag{6.10}$$

where n_{min} is the minimum code length.

In order to analyze the optimum error-free system, we substitute for N and k from (6.4) and (6.5) respectively into (6.7) to get

$$\rho = \left\lfloor \frac{n-1}{\rho(\rho-1)} \right\rfloor, \tag{6.11}$$

which can be rewritten as

$$\rho^2(\rho-1) = n - 1 - (n-1 \mod \rho(\rho-1)). \tag{6.12}$$

Following the same reasoning used to deduce (6.10), we can write the minimum error-free

code length n_{min}^* as

$$n_{min}^* = \rho^2(\rho - 1) + 1. \tag{6.13}$$

Since the RDR algorithm is a greedy algorithm, it can only search for a local optima

of the solution. The optimality in this case is localized to the current construction step

only. Accordingly, the RDR does not guarantee that the achieved solution is the global

optimum across all the different construction steps. This means that the codes designed

using the RDR greedy algorithm (Algorithm 6.4) are generally suboptimal codes. In

order to compare the efficiency of OOCs constructed using RDR with the efficiency of

OOCs constructed using Algorithm 6.1 [85], we define an expansion efficiency factor η

as the ratio between the minimum code length n_{min} and the code length n of the OOC

generated using suboptimal algorithms.

$$\eta = \frac{n_{min}}{n}, \tag{6.14}$$

which means that for an optimum code we have $\eta = 1$, while for a suboptimal code

we have $0 < \eta < 1$. It must be noted, however, that there are no guarantees that for a

general set of parameters $\{n, k, N\}$, an optimal construction method for a (n, w, λ) OOC,

which satisfies the equality condition of the Johnson's bound on N as defined in (6.7),

exists. The existence of optimal constructions have been demonstrated for only a very small subset of the set of different combinations of $\{n, k, N\}$ [88].

6.6 Numerical Results

Figure 6.2 shows the code efficiency for the RDR algorithm η_{RDR} as well as the code efficiency for the greedy algorithm η as a function of code weight k and number of codes N (i.e., code cardinality $|\mathcal{C}|$). Observe from the figure that OOCs designed using the RDR algorithm (Algorithm 6.4) have significantly superior code expansion efficiency over OOCs designed using classical element-by-element greedy algorithm (Algorithm 6.1). Observe also from the figure that the code efficiency is robust against increases in the code cardinality as it does not affect the code efficiency significantly. On the other hand, increasing the code weight rapidly degrades the code efficiency with the RDR being more robust and maintains a superior efficiency across different code weights.

Figures 6.3 and 6.4 are plotted for the case where the system is operating at the maximum capacity under the error free transmission constraints.

Figure 6.3 results demonstrate that codes designed using the RDR algorithm have an asymptotic efficiency of about 0.4 as compared to codes designed using the classical element-by-element greedy algorithm, which have an asymptotic efficiency of about 0.2. This shows that RDR designed codes are expected to constantly outperform classical greedy constructed codes even at asymptotically large values of code weights k.

Figure 6.4 shows the code length n_{RDR} for OOCs designed using the RDR (Algorithm 6.4), the code length n for OOCs designed using Algorithm 6.1, and the minimum

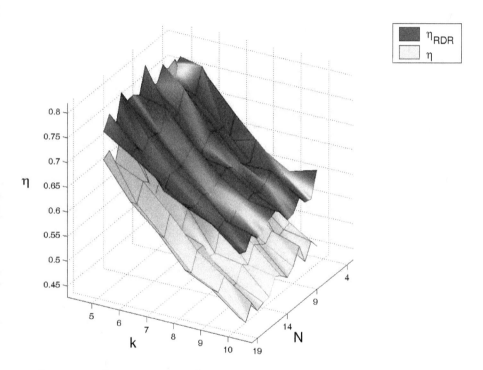

Figure 6.2: Code expansion efficiency factor for RDR greedy and greedy OOCs.

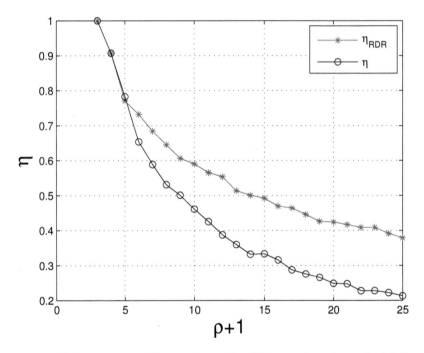

Figure 6.3: Code expansion efficiency factor for RDR greedy and greedy OOCs under error free conditions.

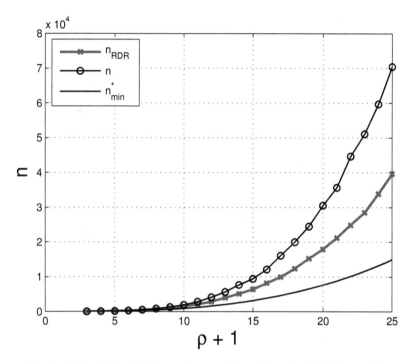

Figure 6.4: Code length for RDR greedy, greedy, and optimal OOCs under error free conditions.

code length n^*_{min} given by (6.13) for the same design parameters under the error free maximum capacity condition (6.4). Clearly the RDR achieves around 30% reduction in the code length n over the classical element-by-element greedy algorithm. This difference becomes more significant as we increase the label space expansion factor ρ (i.e., increase the number of multiplexed users).

6.7 Concluding Remarks

In this chapter, we have proposed a new algorithm for construction of OOCs called Greedy Algorithm with Rejected Delays Reuse (RDR). The algorithm belongs to the family of element-by-element greedy construction algorithms used to generate OOCs. The algorithm incorporates a reuse mechanism for previously rejected smaller delay elements in order to minimize the code length. The codes generated using the RDR algorithm were shown to be generally shorter (have smaller code length n) than those generated using the classical greedy algorithm.

In order to compare the efficiency of the codes generated using the two algorithms we compare their code lengths to the optimum code length using a code expansion efficiency factor η. Numerical results show that OOCs generated using RDR have significantly higher η than those using the classical element-by-element greedy algorithm. It was also found through numerical results that η is largely affected by the increase in the code weight k. However, increasing the number of codes N has negligible effects on η.

Chapter 7

Conclusions and Future Research

We conclude this thesis with a summary of our work, highlights of the major thesis contributions, and some suggestions for future work.

7.1 Summary

The main objectives of this thesis were:

- The increase of sub-wavelength switching granularity in all-optical networks using code division multiplexing.

- The development of enhanced optical code division multiple access schemes.

- The development of mathematical performance models for sub-wavelength code switched mechanisms.

In the following, we present a summary of each of the work chapters. In Chapter 2, we presented an overview of the major transmission methods used in OCDMA communication systems. The overview spanned the different modulation and spreading techniques. It also discussed the different detection methods and the different coding families used in OCDMA systems. In addition, a detailed description accompanied by a mathematical design for a class of OCDMA encoder/decoder devices called Optical Delay Line Correlators is provided. Results in this chapter also appear in [31].

Chapter 3 proposed an architecture for optical networks capable of sub-wavelength switching using OCDMA based on expanding the label space of the GMPLS architecture. The architecture is called Optical Code Labeled GMPLS and employs a code switch capable layer into the GMPLS network to expand the switching granularity of all-optical core switches into the sub-wavelength level. We have evaluated the switching capabilities of the OC-GMPLS using mathematical modeling to derive a proposed performance factor called the label space expansion ratio. Further, we derived the optimum value for the label space expansion ratio under different system constraints. Results of this work have been published in [25, 26].

In Chapter 4, we presented an analytical model for calculating the throughput of the OC-GMPLS core switches. The model incorporates both network and physical layer parameters into the throughput calculation. We have used results from our model to show that there is a set of optimal parameters that maximizes the throughput. The values of these parameters as well as the value of the maximum throughput depend on the provided network and physical layer configurations as well as the required operational constraints. This work also appears in [26, 27].

Chapter 5 proposed a new modulation scheme for OCDMA transmission. The scheme is based on using BPPM to modulate the spread signal chip pulses, hence, the name

Chip-Level Modulated BPPM. The transceiver architecture for implementing the proposed scheme in an all-optical fashion is provided. The proposed scheme enables the receiver to distinguish between the idle and zero transmitting sources within one bit duration. This was not previously possible using OOK methods. The proposed scheme also provides for a better clock recovery over OOK due to the guaranteed transitions within each bit duration as long as the source is active. In order to analyze the proposed system, a general mathematical framework is derived for calculating the bit error rate of any time spread, OOC encoded and correlation detection based OCDMA system. The mathematical model incorporates, for the first time, the average source traffic parameters in the BER calculations. The mathematical framework is applied to the CLM-BPPM to deduce an expression for its BER. The numerical results and comparison with simulations show the high accuracy of the developed model. The results reveal that CLM-BPPM has an error rate that is very close to OOK systems with asymptotically equal values at higher load conditions. Results from this work appear in [28].

Chapter 6 proposed a new method for constructing OOC based on greedy element-by-element recursive construction methods. The proposed algorithm employs a reuse mechanism for previously rejected smaller delay elements in order to produce shorter OOC code lengths. The algorithm is called Greedy Algorithm with Rejected Delays Reuse (RDR). The complexity of the proposed algorithm is analyzed and its resulting code label space expansion efficiency normalized to optimal codes are calculated. The results show that RDR has less complexity than the classical greedy element-by-element methods.

Further, the codes constructed using RDR have an expansion efficiency factor that is asymptotically an order of magnitude higher than that of the classically constructed codes. This work has been published in [29, 30].

7.2   Major Contributions

In conclusion, among the main contributions of this thesis, we proposed:

- An architecture for OCDMA based all-optical sub-wavelength switching in core GMPLS networks. The proposed architecture achieves higher level of flow isolation while maintaining all-optical switching speed at the network core. In order to demonstrate the advantages of our architecture we have derived a switching granularity measure called label space expansion ratio and used it to derive optimal operating parameters under MAI error free condition.

- An analytical model for calculating the throughput of OCDMA-enabled core GMPLS switches. The model incorporates both the physical layer properties through the MAI induced BER as well as network layer parameters such as number of users and average packet length. The model was used to show that there are two major operating regions for the system. Optimal operating parameters which maximize the throughput were numerically deduced under each operating region.

- A new modulation method for OCDMA called CLM-BPPM. The method is based on using BPPM deduced from the bit value to modulate the chip pulses. The method allows for better clock recovery and ability to detect transmitter activity as compared to OOK. It also provides a BER that is very close to OOK with equal asymptotical values at higher user activity.

- A generalized mathematical framework for calculating the multiple access BER in OCDMA networks. The framework can be used to deduce the BER for any time-spread, OOC encoded OCDMA transmission with correlation detection. The accuracy of the mathematical model was compared against simulation models and was shown to achieve very close results.

- A new efficient greedy element-by-element algorithm for constructing shorter optical orthogonal codes. The algorithm was shown to produce smaller code lengths under same design constraints as compared to classical element-by-element greedy methods. The asymptotic computational complexity of the algorithm was deduced and was shown to be smaller than previously proposed algorithms.

7.3 Future Work

We present below a list of research problems that can be investigated as possible extensions for the research work reported in this thesis.

1. The work in Chapter 6 was based on a set theoretic representation of OOC codewords as sets of interdelay elements. The auto and cross correlation properties were mapped into uniqueness properties among one set and across the different sets. It

is worth investigating if a different representation of OOC might lead to better construction methods. An example of a representation that can be investigated is to represent code families as matrices of binary values. Then applying the auto and cross correlation constraints on these matrices one can deduce a set of construction rules for these matrices, using this set of rules in conjunction with a clever matrix search algorithm new construction methods can be established. For this technique to work, it is necessary to find a matrix representation that is unique per family of OOC codes and at the same time can exploit the auto and cross correlation induced constraints to the maximum resulting in minimizing the number of iterations during the matrix search operation.

2. Several variations of the RDR algorithm can be examined in search for a closer to optimal method. In the RDR algorithm codewords at the top of the iteration stack are always assigned shorter inter-delays than those below it. This results in producing a densely occupied area at the start of the code. This in turn will result in higher rejection of inter-delay values at subsequent codewords resulting in increasing their length. This increase in length reflects on the codes at the top of the stack when we reach the terminating stage due to the constraint that all codewords must have the same length. If a more balanced approach like moving through the iterations in a zig-zag manner so that in one round the assignment of inter-delays starts from the top codeword moving down. In the next round it goes in the reverse direction. This way a more balanced assignment of inter-delay values

is achieved, which might result in reducing the code lengths.

3. The throughput of OC-GMPLS derived in Chapter 4 was derived under the assumption of a uniform fiber-wavelength-code assignment combination. In other words, the switch fills up all the fibers and wavelengths with a single flow each. Then, it starts rotating across the different wavelengths over the different fibers assigning a new flow to them one by one. This means that the average number of OCDMA flows on each wavelength are equal. This is not necessarily optimal in terms of the overall throughput. It will be a fruitful point of research to study the different permutations of OCDMA assignment given a certain number of flows to come up with the permutation that maximizes the overall system throughput. The next natural step will be to devise online algorithms that can assign the labels to attain optimality under different loading and channel conditions.

4. The analysis provided in Chapter 5 was carried under ideal optical channel assumptions. It will be interesting to see how the system will perform when optical channel noise and non-linearities are introduced to the model. In particular it will be of great interest to compare the performance of CLM-BPPM with OOK assuming non-ideal optical channel conditions.

5. In optical networks, the error events are very rare events with very low probabilities of values as small as 10^{-12}. In order to directly simulate errors in these networks, simulation time becomes prohibitively large. This problem can be overcome by

deriving mathematical models for the network performance as was done in this thesis. However, mathematical models are not always possible to attain and they require many assumptions that affects their accuracy. Which means that simulation remains an essential tool for analyzing these networks, both to fill in when mathematical modeling is not feasible and to check the accuracy of derived mathematical models. Statistical techniques such as *importance sampling* promise a solution to simulation of rare events. Using such techniques, it would be very useful to try to establish simulation models that can efficiently model optical networks.

Bibliography

[1] A. Kaheel, T. Khattab, A. Mohamed, and H. Alnuweiri. Quality-of-service mechanisms in IP-over-WDM networks. *IEEE Communications Magazine*, 40(12):38–44, December 2002.

[2] B. E. A. Saleh and M. C. Teich. *Fundamentals of Photonics*. John Wiley & Sons, 1991.

[3] K. Iizuka. *Elements of Photonics, Volume II, for Fiber and Integrated Optics*. John Wiley & Sons, 2002.

[4] R. Gagliardi and S. Karp. *Optical Communications—Second Edition*. Wiley Interscience, 1995.

[5] G. Keiser. *Optical Fiber Communications—Third Edition*. McGraw Hill, 2000.

[6] T. Takeda, H. Kojima, and I. Inoue. Optical VPN architecture and mechanisms. In *Proceedings of IEEE 9th Asia-Pacific Conference on Communications APCC'2003*, volume 2, pages 751–755, Penang, Malaysia, 21–24 September 2003.

[7] T. Takeda, H. Kojima, N. Matsuura, and I. Inoue. Resource allocation method for optical VPN. In *Proceedings of IEEE Optical Fiber Communication Conference OFC'2004*, Los Angeles, CA, USA, 22–27 February 2004.

[8] Y. Su, X. Tian, W. Hu, L. Yi, P. Hu, Y. Dong, and H. He. Optical VPN in PON using TDM-FDM signal format. In *Proceedings of IEEE Optical Fiber Communication Conference and the National Fiber Optic Engineers Conference OFC/NFOC'2006*, Anaheim, CA, USA, 5–10 March 2006.

[9] K. N. Oikonomou and R. K. Sinha. Network design and cost analysis of optical VPNs. In *Proceedings of IEEE Optical Fiber Communication Conference and the National Fiber Optic Engineers Conference OFC/NFOC'2006*, Anaheim, CA, USA, 5–10 March 2006.

[10] R. Ramaswami and K. N. Sivarajan. *Optical Networks: A Practical Prespective— Second Edition*. Morgan Kaufmann Publications, 2002.

[11] G. I. Papadimitriou, C. Papazoglou, and A. S. Pomportsis. Optical switching: Switch fabrics, techniques, and architectures. *Journal of Lightwave Technology*, 21(2):384–405, February 2003.

[12] J. M. Elmirghani and H. T. Mouftah. Technologies and architectures for scalable dynamic dense WDM networks. *IEEE Communications Magazine*, 38(2):58–66, February 2000.

[13] T. E. Stern and K. Bala. *Multiwavelength Optical Networks: A Layered Approach.* Prentice Hall PTR, 2000.

[14] T. Fan, P. R. Prucnal, and M. Santoro. Spread spectrum fiber-optic local area network using optical processing. *IEEE Journal of Lightwave Technology,* 4(5):547–554, May 1986.

[15] J. A. Salehi. Code division multiple-access techniques in optical fiber networks—part I: Fundamental principles. *IEEE Transactions on Communications,* 37(8):824–833, August 1989.

[16] F. R. K. Chung, J. A. Salehi, and V. K. Wei. Optical orthogonal code: Design, analysis, and application. *IEEE Transaction on Information Theory,* 35(3):595–604, May 1989.

[17] R. M. Gagliardi, A. J. Mendez, M. R. Dale, and E. Park. Fiber-optic digital video multiplexing using optical CDMA. *IEEE Journal of Lightwave Technology,* 11(1):20–26, January 1993.

[18] L. K. Chen J. G. Zhang and W. C. Kwong. Experimental demonstration of efficient all optical code-division mulltiplexing. *IEE Electronics Letters,* 34(19):1866–1868, September 1998.

[19] Y. H. Lee. Simulation study of different code system division multiple access (DCSDMA) for optical communication. *Microwave and Optical Technology Letters,* 21(4):291–295, May 1999.

[20] T. Dennis, B. Aazhang, and J. F. Young. Demonstration of all-optical CDMA with bipolar codes. In *Proceedings of IEEE Lasers and Electro-Optics Society 10th Annual Meeting LEOS'97*, volume 2, pages 21–22, San Francisco, CA, USA, 10–13 November 1997.

[21] U. N. Griner and S. Arnon. A novel bipolar wavelength-time coding scheme for optical CDMA systems. *IEEE Photonics Technology Letters*, 16(1):332–334, January 2001.

[22] L. L. Jau and Y. H. Lee. Optical code-division multiplexing systems using common-zero codes. *Microwave and Optical Technology Letters*, 39(2):165–167, October 2003.

[23] A. Banerjee, J. Drake, J. P. Lang, and B. Turner. Generalized multiprotocol label switching: An overview of routing and management enhancements. *IEEE Communications Magazine*, 39(1):144–150, January 2001.

[24] D. Awduche and Y. Rekhetert. Mutliprotocol lambda switching: Combining MPLS traffic engineering control with optical crossconnects. *IEEE Communications Magazine*, 39(3):111–116, March 2001.

[25] T. Khattab and H. Alnuweiri. Optical GMPLS networks with code switch capable layer for sub-wavelength switching. In *Proceedings of Global Telecommunications Conference GLOBECOM'04*, volume 3, pages 1786–1792, Dallas, TX, USA, 29 November–3 December 2004.

[26] T. Khattab and H. Alnuweiri. Optical CDMA for all-optical sub-wavelength switching in core GMPLS networks. *Accepted for publication in IEEE Journal on Selected Areas in Communications*, March 2007.

[27] T. Khattab and H. Alnuweiri. Cross-layer throughput analysis for optical code labelled GMPLS networks. In *Proceedings of The IEEE 2nd International Conference on Broadband Networks BROADNETS'05*, volume 1, pages 323–326, Boston, MA, USA, 3–7 October 2005.

[28] T. Khattab and H. Alnuweiri. Chip-level modulated BPPM fiber-optic code division multiple access. *Submitted to IEEE Transactions on Communications*, November 2006.

[29] T. Khattab and H. Alnuweiri. A greedy algorithm for deriving optical orthogonal codes using rejected delays reuse. In *Proceedings of Global Telecommunications Conference GLOBECOM'05*, volume 4, pages 1942–1946, St. Louis, MI, USA, 28 November–2 December 2005.

[30] T. Khattab and H. Alnuweiri. Optical orthogonal code construction using rejected delays reuse for increasing sub-wavelength switching capacity. *IEEE Journal of Lightwave Technology*, 24(9):3280–3287, September 2006.

[31] T. Khattab and H. Alnuweiri. Overview of fiber-optic CDMA communication systems. *Submitted to IEEE Communications Magazine*, December 2006.

[32] A. Stok and E. H. Sargent. Lighting the local area: Optical code-division multiple access and quality of service provisioning. *IEEE Network*, 14(6):42–46, November/December 2000.

[33] A. Stok and E. H. Sargent. System performance comparison of optical CDMA and WDMA in a broadcast local area network. *IEEE Communications Letters*, 6(6):409–411, September 2002.

[34] P. Kamath, J. D. Touch, and J. A. Bannister. The need for media access control in optical CDMA networks. In *Proceedings of IEEE Twenty-third Annual Joint Conference of the IEEE Computer and Communications Societies INFOCOM'2004*, volume 4, pages 2208–2219, Hong Kong, China, 7–11 March 2004.

[35] E. Mutafungwa and S. J. Halme. Analysis of the blocking performance of hybrid OCDM-WDM transport networks. *Microwave and Optical Technology Letters*, 34(1):61–68, July 2002.

[36] M. Murata and K. Kitayama. Ultrafast photonic label switch for asynchronous packets of variable length. In *Proceedings of The Twinty-first Annual Joint Conference of the IEEE Computer and Communications Societies INFOCOM'2002*, volume 1, pages 371–380, New York, NY, USA, 23–27 June 2002.

[37] A. Yariv. *Optical Electronics in Modern Communications—Fifth Edition*. Oxford University Press, 1997.

[38] D. K. Cheng. *Field and Wave Electromagnetics—Second Edition.* Addison-Wesley, 1989.

[39] D. L. LEE. *Electromagnetic Priciples of Integrated Optics—Second Edition.* John Wiley & Sons, 1986.

[40] A. Yariv and P. Yeh. *Photonics: Optical Electronics in Modern Communications— Sixth Edition.* Oxford University Press, 2007.

[41] C. Xu, X. Liu, and X. Wei. Differential phase-shift keying for high spectral efficiency optical transmissions. *Journal of Selected Topics in Quantum Electronics*, 10(2):281–293, March/April 2004.

[42] D. S. Ly-Gagnon, S. Tsukamoto, K. Katoh, and K. Kikuchi. Coherent detection of optical quadrature phase-shift keying signals with carrier phase estimation. *Journal of Lightwave Technology*, 24(1):12–21, January 2006.

[43] T. Kawanishi, T. Sakamoto, S. Shinada, M. Izutsu, T. Fujita, K. Higuma, and J. Ichikawa. 10 Gbit/s FSK transmission over 130 Km SMF using group delay compensated balance detection. In *Technical Digest of Optical Fiber Communication Conference OFC/NFOEC'05*, volume 2, Anaheim, CA, USA, 6–11 March 2005.

[44] F. Khansefid, R. Gagliardi, and H. Taylor. Performance analysis of code division multiple access techniques in fiber optics with On-Off and PPM pulsed signaling. In *Proceedings of IEEE Military Communications Conference MILCOM'90*, volume 3, pages 909–915, Monterey, CA, USA, 30 September–3 October 1990.

[45] H. M. H. Shalaby. Performance analysis of optical CDMA communication systems with PPM signaling. In *Proceedings of IEEE Global Telecommunications Conference GLOBECOM'93*, volume 3, pages 1901–1905, Houston, TX, USA, 29 November–2 December 1993.

[46] H. M. H. Shalaby. Performance analysis of optical synchronous CDMA communication systems with PPM signaling. *IEEE Transactions on Communications*, 43(2/3/4):624–634, Fibruary/March/April 1995.

[47] H. M. H. Shalaby. Maximum achievable number of users in optical PPM-CDMA. In *Proceedings of IEEE International Symposium on Information Theory*, page 477, Ulm, Germany, 29 Jun–4 July 1997.

[48] H. M. H. Shalaby. A performance analysis of optical overlapping PPM-CDMA communication systems. *IEEE Network*, 17(3):426–433, March 1999.

[49] H. M. H. Shalaby. Maximum achievable number of users in optical PPM-CDMA local area networks. *IEEE Journal of Lightwave Technology*, 18(9):1187–1196, September 2001.

[50] C. Argon and S. W. McLaughlin. Optical OOK-CDMA and PPM-CDMA systems with turbo product codes. *Journal of Lightwave Technology*, 20(9):1653–1663, September 2002.

[51] K. S. Kim, D. M. Marom, L. B. Milstein, and Y. Fainman. Hybrid pulse position modulation/ultrashort light pulse code-division multiple-access systems—part

I: Fundamental analysis. *IEEE Transactions on Communications*, 50(12):2018–2031, December 2002.

[52] K. S. Kim, D. M. Marom, L. B. Milstein, and Y. Fainman. Hybrid pulse position modulation/ultrashort light pulse code-division multiple-access systems—part II: Timespace processor and modified schemes. *IEEE Transactions on Communications*, 51(7):1135–1148, July 2003.

[53] M. Azizoglu, J. A. Salehi, and Y. Li. Opical CDMA via temporal codes. *IEEE Transactions on Communications*, 40(7):1162–1170, July 1992.

[54] M. Kavehrad and D. Zaccarh. Optical code-division-multiplexed systems based on spectral encoding of noncoherent sources. *IEEE Journal of Lightwave Technology*, 13(3):534–545, March 1995.

[55] K. Yu, J. Shin, and N. Park. Wavelength-time spreading optical CDMA system using wavelength multiplexers and mirrored fiber delay lines. *IEEE Photonics Technology Letters*, 12(9):1278–1280, September 2000.

[56] B. Ni. The performances of optical code-division multiple access communication systems. ProQuest Thesis Online UMI Number: 3185704, August 2005.

[57] K. Murugesan and V. C. Ravichandran. Evaluation of new codes for spectral-amplitude coding optical code-division multipleaccess communication systems. *SPIE Optical Engineering*, 43(4):911–917, April 2004.

[58] P. R. Prucnal, editor. *Optical Code Division Multiple Access*. CRC Press, 2006.

[59] L. R. Chen, S. D. Benjamin, P. W. E. Smith, and J. E. Sipe. Applications of ultra-short pulse propagation in bragg gratings for wavelength-division multiplexing and code-division multiple access. *IEEE Journal Of Quantum Electronics*, 34(11):2117–2129, November 1998.

[60] R. M. H. Yim, J. Bajcsy, and L. R. Chen. A new family of 2-D wavelength-time codes for optical CDMA with differential detection. *IEEE Photonics Technology Letters*, 15(1):165–167, January 2003.

[61] J. G. Zhang, W. C. Kwongz, and A. B. Sharma. Effective design of optical fiber code-division multiple access networks using the modified prime codes and optical processing. In *Proceedings of International Conference on Communication Technology WCC-ICCT'2000*, volume 1, pages 392–397, Beijin, China, 21–25 January 2000.

[62] Y. Shimura, H. Yashima, and J. Suzuki. An analysis of the cross-correlation properties of prime code and bit error rate in an optical CDMA system. *Electronics and Communications in Japan, Part 3*, 83(9):94–103, September 2000.

[63] S. Martirosyan and A. J. H. Vinck. A construction for optical orthogonal codes with correlation 1. *IEICE Transactions on Fundamentals of Electronics*, E85-A(1):269–272, January 2002.

[64] A. Keshavarzian and J. A. Salehi. Multiple-shift code acquisition of optical orthogonal codes in optical CDMA systems. *IEEE Transactions on Communications*, 53(4):687–697, April 2005.

[65] A. D. Neto and E. Moschim. Some optical orthogonal codes for asynchronous CDMA systems. In *Proceedings of IEEE Global Telecommunications Conference GLOBE-COM'2002*, volume 3, pages 2065–2068, Taipei, Taiwan, China, 17–21 November 2002.

[66] J. A. Salehi and C. A. Brackett. Code division multiple-access techniques in optical fiber networks—part II: Performance analysis. *IEEE Transactions on Communications*, 37(8):834–842, August 1989.

[67] S. Zahedi and J. A. Salehi. Analytical comparison of various fiber-optic CDMA receiver structures. *IEEE Journal of Lightwave Technology*, 18(12):1718–1727, December 2000.

[68] H. M. H. Shalaby. Chip-level detection in optical code division multiple access. *IEEE Journal of Lightwave Technology*, 16(6):1077–1087, June 1998.

[69] A. Iwata, K. Kamakura, and I. Sasase. A synchronization method deselecting candidate positions with chip level detection for optical CDMA. *Electronics and Communications in Japan, Part 1*, 88(3):21–32, March 2005.

[70] G. P. Agrawal. *Fiber-Optic Communication Systems—Third Edition*. Wiley Interscience, 2002.

[71] A. J. Viterbi. *CDMA Priciples of Spread Spectrum Communication.* Addison-Wesley Longman Inc., 1995.

[72] L. Zhang and J. Tang. Label-switching architecture for IP traffic over WDM networks. *IEE Proceedings on Communications*, 147(5):269–276, May 2001.

[73] Y. G. Wen, Y. Zhang, and L. K. Chen. On architecture and limitation of optical multiprotocol label switching (MPLS) networks using optical-orthogonal-code (OOC)/wavelength label. *Optical Fiber Technology*, 8(1):43–70, January 2002.

[74] M. Murata and K. Kitayama. A prespective on photonic multiprotocol label switching. *IEEE Network*, 15(4):56–63, July/August 2001.

[75] A. Kaheel and H. Alnuweiri. A strict priority scheme for quality-of-service provisioning in optical burst switching networks. In *Proceedings of IEEE Symposium on Computers and Communications ISCC'03*, volume 1, pages 16–21, Kemer, Antalya, Turkey, 30 June–3 July 2003.

[76] K. Kamakura, O. Kabranov, D. Makrakis, and I. Sasase. OBS networks using optical code division multiple access techniques. In *Proceedings of IEEE Twenty-third Annual Joint Conference of the IEEE Computer and Communications Societies INFOCOM'2004*, volume 3, pages 1725–1729, Paris, France, 20–24 June 2004.

[77] E. E. Mannie. Generalized multi-protocol label switching (GMPLS) architecture. available online: "http://www.ietf.org/rfc/rfc3945.txt", October 2004.

[78] A. Mohamed, A. Kaheel, T. Khattab, and H. Alnuweiri. Evaluation of optical packet switch as edge device using OPNET modeler. available online: "http://www.opnet.com/opnetwork2002", August 2002.

[79] T. Khattab, A. Mohamed, A. Kaheel, and H. Alnuweiri. Optical packet switching with packet aggregation. In *Proceedings of International Conference on Software, Telecommunications and Computer Networks SoftCOM'02*, volume 1, pages 741–746, Split, Dubrovnik, Croitia/Ancona, Bari, Italy, 9–12 October 2002.

[80] International Telecommunication Union. Spectral grids for WDM applications: DWDM frequency grid. available online: "http://www.itu.int/rec/T-REC-G.694.1/en", June 2002.

[81] D. Raychaudhuri. Performance analysis of random access packet-switched code division multiple access systems. *IEEE Transactions on Communications*, 29(6):895–901, June 1981.

[82] R. Gagliardi and S. Karp. *Optical Communications*. John Wiley & Sons, 1975.

[83] G. Jacobsen. *Noise in Digital Optical Transmission Systems*. Artech House, 1994.

[84] G. Einarsson. *Principles of Lightwave Communications*. John Wiley & Sons, 1996.

[85] C. Argon and R. Ergl. Optical CDMA via shortened orthogonal codes based on extended sets. *Optics Communications*, 116(5):326–330, May 1995.

[86] Jr. M. Hall. *Combinatorial Theory—Second Edition*. Wiley Interscience, 1986.

[87] I. Anderson. *Combinatorial Designs: Construction Methods.* Ellis Horwood Limited, 1990.

[88] I. B. Djordjevic and B. Vasic. Combinatorial constructions of optical orthogonal codes for OCDMA systems. *IEEE Communications Letters*, 8(6):391–393, January 2004.

Appendix A

Proof of $p_r \leq 1$ in Equation 4.1

In this Appendix, we provide a mathematical proof that a *non-trivial OOC*, which satisfies the orthogonality condition (3.3) satisfies also the following condition:

$$p_r = \frac{k^2}{n} \leq 1 .$$

Proof. Starting from the orthogonality condition, we have

$$N \leq \left\lfloor \frac{n-1}{k(k-1)} \right\rfloor . \qquad (A.1)$$

Therefore, we can write

$$N \leq \left\lfloor \frac{n-1}{k(k-1)} \right\rfloor \leq \frac{n-1}{k(k-1)} , \qquad (A.2)$$

which can be rearranged to give

$$N \cdot k(k-1) \leq n-1 . \qquad (A.3)$$

For a *non-trivial OOC*, $N > 1$. However, for a given n, the largest value of k in (A.3) is achieved by choosing the smallest possible value of N, which is achieved by choosing $N = 2$. Hence,

$$2k(k-1) \leq n - 1 , \tag{A.4}$$

which can be rewritten as

$$k^2 + (k-1)^2 \leq n . \tag{A.5}$$

Since

$$(k-1)^2 \geq 0 ,$$

(A.5) can be rewritten as

$$k^2 \leq n , \tag{A.6}$$

Substituting (A.6) in (4.1) we get

$$p_r = \frac{k^2}{n} \leq 1 .$$

□

Appendix B

Derivation of Equation 5.28

In Chapter 5, we gave a simplified mathematical representation of the BER for a CLM-BPPM using absolutely timed detection and '1'-value first bit conflict resolution by (5.28). In the following, we derive this equation in details.

Starting from the right hand side (RHS) of (5.27), we have

$$
\text{RHS} = (1 - P_A P_1) \overbrace{\sum_{l=h}^{N_u-1} \sum_{m=0}^{N_u-1-l} P(l,m)}^{R_1} + (1 - P_A) \underbrace{\sum_{l=0}^{h-1} \sum_{m=h}^{N_u-1-l} P(l,m)\phi(l,m)}_{R_2} . \tag{B.1}
$$

The first term ($R1$) can be written as

$$
\begin{aligned}
R_1 &= \sum_{l=h}^{N_u-1} \binom{N_u-1}{l} q_1^l (1-q_1-q_0)^{N_u-1-l} \sum_{m=0}^{N_u-1-l} \binom{N_u-1-l}{m} q_0^m (1-q_1-q_0)^{-m} \\
&= \sum_{l=h}^{N_u-1} \binom{N_u-1}{l} q_1^l (1-q_1-q_0)^{N_u-1-l} \sum_{m=0}^{N_u-1-l} \binom{N_u-1-l}{m} \left(\frac{q_0}{1-q_1-q_0}\right)^m \\
&= \sum_{l=h}^{N_u-1} \binom{N_u-1}{l} q_1^l (1-q_1-q_0)^{N_u-1-l} \left(1 + \frac{q_0}{1-q_1-q_0}\right)^{N_u-1-l} \\
&= \sum_{l=h}^{N_u-1} \underbrace{\binom{N_u-1}{l} q_1^l (1-q_1)^{N_u-1-l}}_{P(l)} ,
\end{aligned}
$$

which reduces to

$$R_1 = 1 - \sum_{l=0}^{h-1} P(l) \,, \tag{B.2}$$

where $P(l)$ is the binomial distribution with parameters q_1, $N_u - 1$ and l.

Considering the second term $(R2)$, we can rewrite it as

$$R_2 = \sum_{l=0}^{h-1} \binom{N_u - 1}{l} q_1^l (1 - q_1 - q_0)^{N_u - 1 - l} S(l) \phi(l) \,, \tag{B.3}$$

where

$$
\begin{aligned}
S(l) &= \sum_{m=h}^{N_u - 1 - l} \binom{N_u - 1 - l}{m} \left(\frac{q_0}{1 - q_1 - q_0} \right)^m \\
&= \left(\frac{1 - q_1}{1 - q_1 - q_0} \right)^{N_u - 1 - l} - \sum_{m=0}^{h-1} \binom{N_u - 1 - l}{m} \left(\frac{q_0}{1 - q_1 - q_0} \right)^m \,,
\end{aligned} \tag{B.4}
$$

and

$$
\phi(l) = \begin{cases} 1 & l \le N_u - 1 - h \\[2mm] 0 & \text{otherwise.} \end{cases}
$$

is $\phi(l, m)$ rewritten as a function that depends on l only.

Substituting (B.4) in (B.3), we get

$$R_2 = \sum_{l=0}^{h-1} \binom{N_u - 1}{l} q_1^l (1 - q_1)^{N_u - 1 - l} \phi(l) - \sum_{m=0}^{h-1} \binom{N_u - 1}{l, m} q_1^l q_0^m (1 - q_1 - q_0)^{N_u - 1 - l - m} \phi(l)$$

$$= \sum_{l=0}^{h-1} P(l)\phi(l) - \sum_{l=0}^{h-1} \sum_{m=0}^{h-1} P(l, m)\phi(l) \ . \tag{B.5}$$

Substituting (B.2) and (B.5) in (B.1) results in

$$\mathrm{RHS} = (1 - P_A P_1) \left(1 - \sum_{l=0}^{h-1} P(l) \right) + (1 - P_A) \left(\sum_{l=0}^{h-1} P(l)\phi(l) - \sum_{l=0}^{h-1} \sum_{m=0}^{h-1} P(l, m)\phi(l) \right)$$

$$= 1 - P_A P_1 + \sum_{l=0}^{h-1} P(l) \left[(1 - P_A)\phi(l) - (1 - P_A P_1) \right] - (1 - P_A) \left(\sum_{l=0}^{h-1} \sum_{m=0}^{h-1} P(l, m)\phi(l) \right) \ .$$

Appendix C

Asymptotic Computation Time for

the RDR Algorithm

In this appendix, we provide an asymptotic computational complexity analysis for the RDR element-by-element greedy construction algorithm described in details in Chapter 6. The RDR algorithm performs its operations on the set $\mathbb{T}_{ext}$, which is composed of unique elements $1 \leq e_{ij} \leq n$. Hence, the algorithm tests at most n elements (running at most n iterations). For each iteration, a new interdelay element is tested for addition to code number i at delay element number j in $\mathbb{T}$ at step (i,j), where $1 \leq i \leq N$ and $1 \leq j \leq k$. Assuming that the algorithm will perform equal number of iterations in each step (i,j) to reach the accepted interdelay element, the total number of iterations per step will be

$$\frac{n}{N \times k} \ .$$

In each iteration, the algorithm performs linear time operations (comparisons, additions, etc.) on, at most, all the elements of $\mathbb{T}_{ext}^{(i,j)}$, where $\mathbb{T}_{ext}^{(i,j)}$ is the partial set of $\mathbb{T}_{ext}$ composed at step (i,j) of the algorithm. Taking the worst case scenario when $i = N$ (an element is tested for addition to code number N), the total number of operations performed on all the elements of $\mathbb{T}_{ext}^{(i,j)}$ per iteration per step is given by:

$$c_o \times N \times j(j-1) \, ,$$

where c_o is an arbitrary constant. Accordingly, the maximum total number of operations N_o performed by the algorithm is given by:

$$
\begin{aligned}
N_o &= \sum_{i=1}^{N} \sum_{j=1}^{k} \frac{n}{N \times k} \times c_o \times N \times j(j-1) \\
&= c_o \frac{n \times N}{k} \sum_{j=1}^{k} j(j-1) \\
&= c_o \frac{n \times N}{k} \cdot \frac{(k-1)k(k+1)}{3} \\
&= \frac{c_o}{3} \times n \times N(k-1)(k+1) \, .
\end{aligned}
$$

This means that the algorithm has $O(n \times N \times k^2)$ computation complexity. However, we know from Johnson's inequality (proven in Theorem 6.5), that

$$N \leq \frac{n-1}{k(k-1)} \, .$$

Therefore,

$$O(N) \leq O(\frac{n}{k^2}) \, .$$

Consequently, the computation complexity of the algorithm is

$$O(n \times \frac{n}{k^2} \times k^2) = O(n^2) \ .$$

www.ingramcontent.com/pod-product-compliance
Lightning Source LLC
LaVergne TN
LVHW022311060326
832902LV00020B/3390